AF807070

World-Systems Evolution and Global Futures

Series Editors

Christopher Chase-Dunn, University of California, Riverside, CA, USA

Barry K. Gills, Political and Economic Studies, University of Helsinki, Helsinki, Finland

Leonid E. Grinin, National Research University Higher School of Economics, Moscow, Russia

Andrey V. Korotayev, National Research University Higher School of Economics, Moscow, Russia

This series seeks to promote understanding of large-scale and long-term processes of social change, in particular the many facets and implications of globalization. It critically explores the factors that affect the historical formation and current evolution of social systems, on both the regional and global level. Processes and factors that are examined include economies, technologies, geopolitics, institutions, conflicts, demographic trends, climate change, global culture, social movements, global inequalities, etc. Building on world-systems analysis, the series addresses topics such as globalization from historical and comparative perspectives, trends in global inequalities, core-periphery relations and the rise and fall of hegemonic core states, transnational institutions, and the long-term energy transition. This ambitious interdisciplinary and international series presents cutting-edge research by social scientists who study whole human systems and is relevant for all readers interested in systems approaches to the emerging world society, especially historians, political scientists, economists, sociologists, geographers and anthropologists.

This book series is indexed in Scopus.

All titles in this series are peer-reviewed.

Tessaleno Devezas • João Leitão •
Askar Sarygulov • David J. LePoire •
Bulat Khusainov
Editors

Global Energy Transition and Sustainable Development Challenges, Vol. 1

Models and Regions

Springer

Editors
Tessaleno Devezas
Universidade Atlântica
Barcarena, Portugal

João Leitão
Faculty of Human and Social Sciences
University of Beira Interior
Covilhã, Portugal

Askar Sarygulov
CIRETEC-GT
Peter the Great St. Petersburg Polytechnic University
Saint Petersburg, Russia

David J. LePoire
Argonne National Laboratory
Lemont, IL, USA

Bulat Khusainov
Zhetysu University
Taldykorgan, Kazakhstan

ISSN 2522-0985 ISSN 2522-0993 (electronic)
World-Systems Evolution and Global Futures
ISBN 978-3-031-67582-9 ISBN 978-3-031-67583-6 (eBook)
https://doi.org/10.1007/978-3-031-67583-6

© The Editor(s) (if applicable) and The Author(s), under exclusive license to Springer Nature Switzerland AG 2024
This work is subject to copyright. All rights are solely and exclusively licensed by the Publisher, whether the whole or part of the material is concerned, specifically the rights of translation, reprinting, reuse of illustrations, recitation, broadcasting, reproduction on microfilms or in any other physical way, and transmission or information storage and retrieval, electronic adaptation, computer software, or by similar or dissimilar methodology now known or hereafter developed.
The use of general descriptive names, registered names, trademarks, service marks, etc. in this publication does not imply, even in the absence of a specific statement, that such names are exempt from the relevant protective laws and regulations and therefore free for general use.
The publisher, the authors and the editors are safe to assume that the advice and information in this book are believed to be true and accurate at the date of publication. Neither the publisher nor the authors or the editors give a warranty, expressed or implied, with respect to the material contained herein or for any errors or omissions that may have been made. The publisher remains neutral with regard to jurisdictional claims in published maps and institutional affiliations.

This Springer imprint is published by the registered company Springer Nature Switzerland AG
The registered company address is: Gewerbestrasse 11, 6330 Cham, Switzerland

If disposing of this product, please recycle the paper.

Introduction

This two-volume book intends to be a follow-up of a previous two-volume book titled *Global Challenges of Climate Change*, published in the series *World Systems Evolution and Global Futures*.

As is well known, the phenomenon of energy transition has been the subject of widespread discussion since August 1990 and has been on the world development agenda for more than 30 years. Therefore, the question quite naturally arises of how much the global balance of energy consumption has grown in favor of renewable energy sources (RES), but also how much the public consciousness has really changed over this period of time. There is no doubt that one of the key factors that influenced the change in people's consciousness and their awareness of the need for a more limited use of hydrocarbon resources was global warming and climate change. Indeed, we are witnessing changes in weather conditions, rising sea levels, and destructive climate events such as hurricanes, forest fires, droughts, floods, etc. which have become more frequent. Many countries around the world, reacting to these changes, have developed long-term plans to actively replace fossil fuels—gas, oil, and coal with renewable energy sources, mainly solar and wind. However, the not very high replacement rates observed in the global energy sector over the past 30 years raise the question of how far the decarbonization scenarios and models being implemented by many countries bring us closer to the ultimate goal of creating an economy with a low-carbon footprint.

Today we are witnessing a number of contradictory trends in global energy. On the one hand, during this period the worldwide energy consumption grew about 58% energy (oil grew 55%, gas 71%, and coal 32%). And this growth is largely due to significant investment in the oil and gas sector, which between 1990 and 2018 fluctuated between 0.3% and 1.1% of global GDP. Coal still remains an important energy resource and accounts for more than 32% of global final energy consumption. This high share is largely due to the fact that fast-growing large economies, primarily China and India, which account for two-thirds of global coal consumption, view this resource as the most accessible. On the other hand, in the period from 2010 to 2020 the shares of solar and wind energy in total generation increased by 5 and 2.5 times, respectively. This became possible due to large volumes of investment in this energy segment: the total volume of new investments in these two renewable energy resources during the period under review, according to various sources, ranged from 2.6 to 4.8 trillion US$. The annual increase in installed capacity in solar and wind generation in 2020–2021 was about 235 GW. The latest IEA forecasts show higher rates of capacity growth: from 290 GW in 2021 to 1200 GW in 2030. This means, with the ratio remaining today, solar and wind generation will account for about 970 GW of annual capacity growth and its growth rate will be four times higher than in the early 2020s. Along with renewable energy, electrification of transport is another key point of investment for financial resources in order to decarbonize the economy. In 2021, the global fleet of electric vehicles exceeded 17 million units and by 2030, according to IEA estimates, the global fleet will approach 350 million units. Such ambitious plans to increase capacity in renewable energy and electrification of transport raise the issue of resource provision, first of all, regarding the materials necessary for the production of turbines, solar panels, and electric batteries. The positions of skeptics of the energy transition are only strengthened against the backdrop of data on the dynamics of changes in the share of low-carbon resources (nuclear, hydropower, bioenergy, solar, and wind) in the primary global energy balance for 1990–2021. This share has increased over 30 years from 13.12% to 17.72%, i.e. by 4.6% points, despite significant volumes of investment.

Considering that in the modern world there are different approaches of energy policy, this book aims to give a more complete understanding of the driving forces and factors that can have a significant impact on the pace and speed of the global energy transition. Among such driving forces and factors, we highlight, first of all, the unevenness of regional development. This unevenness also predetermines the emergence of different models and scenarios of the energy transition. Materials and technologies are both drivers and constraints of energy transitions.

In this first volume of this book, it is presented a collection of chapters focusing on the development of models and discussions on the unevenness of regional and country development, which content is resumed in the next lines:

Chapter 1: Tausch—*The Greening of the Global South? Analysing World Values Survey and European Values Survey Data on Environmental Movements in 88 countries and territories from 2017-2021*

This study focuses on the analysis of public opinion survey data on trust in and participation in environmental movements from the World Values Survey and the European Values Survey, covering 88 countries and territories from 2017 onward. Are we witnessing a greening of the Global South? Most of the existing, highly influential studies on this topic are based on earlier waves of the World Values Survey and the European Values Survey, with only a limited number of countries from the Global South covered in the surveys. Our various steps of multivariate analysis seem to suggest that, in today's world, it is rather the Global South that is "greening" and where environmental movements enjoy the highest levels of participation and confidence. We can also show that, holding human development indices constant, this greening of the Global South has overall positive effects on the socio-economic development of the countries concerned.

Chapter 2: Mendonça—*Creative Regulatory Environments: Sandboxes at the Intersection of Regulation and Innovation in an Era of Transition*

The "Regulatory Sandboxes" are positioned as policy instruments that act at the interface of regulation and innovation processes, resulting in "Creative Regulatory Environments." Sandboxes are a relatively new weapon in the armory of national administrative bodies tasked with regulating markets, overseeing industries, and monitoring products or infrastructures. They have various benefits, including reduced uncertainty for operators and facilitating learning by regulators, and their implementation might result in a dynamic (re)mix(ed) economy.

Chapter 3: Moutinho, Leitão, and Henriques—*Examining the Behavior of the Kuznets Curve under Economic Policy Uncertainty Conditions Using the Stochastic Frontier Analysis: Evidence from 11 Selected Countries*

The behavior of the Kuznets Curve (EKC) under economic policy uncertainty is investigated by making use of a Stochastic Frontier Analysis. Using data from 11 G7 and BRIC countries, the empirical evidence obtained reveals that, on the one hand, following the EKC formulation, an Inverted U Shape configuration is confirmed for the Stochastic Frontier, concerning the relationship between total resource rents and CO2 emissions per capita. In the case of the link between the uncertain economic policy index (EPU) and CO2 emissions per capita, however, the identical configuration is verified for one border variable model. It is advised that new economic policies that are in accordance with clean, greener energy and energy security need to be developed in the face of economic policy uncertainty.

Chapter 4: Kolpakov, Maimekov—*Carbon Intensity of Products under the EU CBAM: Assessment of Russia's Competitiveness Based on Inter-Country Analysis*

The EU's Carbon Border Adjustment Mechanism (CBAM) aims to expand the concept of competitiveness of products in cross-country trade by adding a carbon footprint metric. This chapter is devoted to this mechanism. The authors show that estimating the carbon intensity of products that will be impacted by CBAM is possible on the basis of IEA data and national inventories of anthropogenic emissions. In addition, it is shown that most Russian goods are characterized by a carbon footprint that is not higher than the weighted average level on the market, with the exception of goods such as construction materials: cement, lime, glass, which are not exported to EU countries.

Chapter 5: Khusainov, Nusupov, Kaimodina, Temirbayev—*Inclusive Growth of National Economies and a New Energy Paradigm*

The concept of inclusive growth is part of a new multi-dimensional initiative by the OECD to identify new approaches to economic challenges. This chapter analyzes the impact of energy consumption on the inclusive growth of European and Asia-Pacific economies in the context of the new energy transition paradigm. The authors understand inclusive growth as equitable growth for all segments of society, which aims to reduce growing income heterogeneity. This is one of the first studies to assess the impact of the consumption of different types of energy resources, reflecting the problems of the energy transition, on the inclusive growth of the economies of Europe, Central Asia, and the Asia-Pacific region and covers the economies of 55 countries over a 20-year retrospective period. The authors show that the main hypothesis of the study about the impact of energy consumption on the inclusive growth of national economies and the presence of multidirectional influence of six types of energy resources on the inclusive growth of the economies of these countries has been quantitatively confirmed.

Chapter 6: Buktukov, Gumennikov, Moldabaeva—*Solutions to the Problems of Transition to Green Energy in Kazakhstan*

Coal energy brings about significant environmental damage by occupying vast areas of agricultural land for waste dumps and ash landfills. This chapter is devoted to finding solutions to the problems of the coal industry through the use of more complex energy technologies. In particular, it is shown that the use of more complex technologies can reduce the use of very rare and expensive metals, as well as reduce unreasonably high energy costs for the creation of solar and wind energy.

Chapter 7: Moutinho; Oliveira, Oliveira, Vaz—*How Natural Resources, Corruption, Globalization and CO2 Emissions Affect the Economic Growth in Europe?*

This study looks at the influence of economic, political, social, and environmental variables on economic development in European Union member countries. A two-step econometric strategy is built using scoreboard data for the 27 European Union Member States from 1995 to 2019. First, the static analysis demonstrates that all coefficients linked with the identified economic development factors are statistically significant. Second, the estimators' long-term findings reveal that the variables CO2 emissions and the corruption perception index have a positive influence on the economic growth of various nations. In contrast, the variables Total Natural Resource Income and Jure Political Globalization Index have a negative impact. This research is critical for developing well-informed strategies for globalization integration across several dimensions.

Chapter 8: Yegorov—*The Problem of Regional Energy Scarcity in Europe*

Global energy resources are characterized by substantial spatial heterogeneity across continents, countries, and regions. Even in globalized world geographic approach to modeling was important, for example, for natural gas. The failure of globalization and coming regionalization put a pressure for some regions, Europe, in particular. It needs ambitious fast transition to green energy. But even here we observe internal regional heterogeneity related to possibilities in regional production of water, solar, and wind energy. Stochastic supply calls for building powerful high-voltage network to balance regional demand with supply, and this requires higher integration inside EU. This chapter discusses this issue from theoretical perspective of energy security after green transition, considering both physical and economic issues. The influence of geopolitics on natural gas scarcity is also discussed.

Chapter 9: Cotton—*Policy, Market, and Skills Barriers to Heat Pump Deployment in the United Kingdom*

UK domestic heat decarbonization strategy is undergoing rapid policy change. The lag in energy efficiency performance improvements in domestic hot water and space heating—when compared to the electricity sector—is a key barrier to achieving national net zero policy targets. This paper outlines the key issues raised in domestic space heating and hot water decarbonization in the UK, with a specific focus on the uptake of air and ground-source heat pumps—a major component of government strategy for future homes. The key policy, market, and skills barriers to the rapid uptake of heat pumps are assessed, including issues of retrofitting older housing stock, encouraging market uptake of subsidized technologies, and retraining fossil-fuel intensive hot water and heating systems professionals to encourage consumer confidence. The key policy recommendations from this analysis are to reframe heat policy through a more holistic approach to both "supply push" and "demand pull" in the heat pump market—not only ensuring new building standards, but improving funding for skills, quality control, and compliance in the sector to reduce costs, speed up market growth, improve consumer confidence, and ensure a "just transition" in the heating sector.

Chapter 10: Simões—*Paving the Way for Sustainable Development in Small Island Developing States: Insights from Cape Verde's Energy Transitions*

This chapter examines the outlook for energy transitions in Cape Verde, a small island developing state located in the Atlantic Ocean, off the western coast of Africa. It explores diverse dimensions, including energy profiles, national policies, institutional frameworks, and emerging challenges and opportunities. The primary research question we pose is: what challenges does Cape Verde face in its energy transitions, and how can these energy transitions foster overall socio-economic development? Additionally, the chapter discusses the role of international cooperation. It draws on various sources, including scientific and gray literature, and official national documents, such as development plans and reports. By highlighting the challenges of national transitions in the context of Cape Verde, this chapter contributes to a more comprehensive understanding of energy transitions and sustainable development in small island developing states.

It should be highlighted that Chaps. 4 and 5 of this book were presented at the First International Scientific Conference "*A New Path to Decarbonization of the Economy* (KZGT)," which was held on November 8–10, 2023, at the Al Farabi Kazakh National Research University (Almaty, The Republic of Kazakhstan), supported by the projects: "*Technological challenges and socioeconomic transformation in the context of green transitions*" (Chap. 4), financed by the Ministry of Science and Higher Education of the Russian Federation (Grant Agreement No. 075-15-2022-1136, dated 1 July 2022) and "*The development imperatives of Kazakhstan in Eurasian integration against the backdrop of challenges and fault lines in the global architecture,*" (Chap. 5) funded by the Committee of Science of the Ministry of Science and Higher Education of the Republic of Kazakhstan (Grant No. BR21882364).

Adding to the previous, two chapters e.g. Chaps. 3 and 7, were developed in the scope of the research team associated with the PRIMA project: "*Improving the*

MEDiterranean irrigation system and water supply for small farmers by providing efficient, low-cost and nature-based technologies and practices," with reference PRIMA/0008/2020, and the DOI: 10.54499/PRIMA/0008/2020.

Lisbon, Fundão, Saint Petersburg, Lemont, Almaty, June 2024

The Editors

Contents

Chapter 1
The Greening of the Global South? Analysing World Values Survey and European Values Survey Data on Environmental Movements in 88 Countries and Territories from 2017 to 2021

Arno Tausch

Abstract This study focuses on the analysis of public opinion survey data on trust in and participation in environmental movements from the World Values Survey and the European Values Survey, covering 88 countries and territories from 2017 onwards. Are we witnessing a greening of the Global South? Most of the existing, highly influential studies on this topic are based on earlier waves of the World Values Survey and the European Values Survey, with only a limited number of countries from the Global South covered in the surveys.

Our various steps of multivariate analysis seem to suggest that, in today's world, it is rather the Global South that is "greening" and where environmental movements enjoy the highest levels of participation and confidence. We can also show that, holding human development indices constant, this greening of the Global South has overall positive effects on the socio-economic development of the countries concerned.

1.1 Introduction

This study focuses on the analysis of public opinion survey data on trust in and participation in environmental movements from the World Values Survey and the European Values Survey, covering 88 countries and territories from 2017 onwards. Are we witnessing a greening of the Global South? Most of the existing, highly influential studies on this topic are based on earlier waves of the World Values Survey and the European Values Survey, with only a limited number of countries from the Global South included in the surveys.

A. Tausch (✉)
Visiting Professor of Political Studies and Governance, University of the Free State, Bloemfontein, South Africa

© The Author(s), under exclusive license to Springer Nature Switzerland AG 2024

T. Devezas et al. (eds.), *Global Energy Transition and Sustainable Development Challenges, Vol. 1*, World-Systems Evolution and Global Futures,
https://doi.org/10.1007/978-3-031-67583-6_1

The importance of the issues analysed here cannot be underestimated. Brand (2015, 2023), Brand and Lang (2015) argue that in contrast to the concept and strategy of sustainable development in the 1990s, a green economy now appears to be attractive to relevant socio-economic actors. Technologies are available to develop renewable energy sources or electronic engines for cars, and microelectronics play a much more important role today than 20 years ago. And there is another dynamic, namely the current crisis, the main cause of which is an enormous amount of over-accumulated capital looking for new investment opportunities. Brand (2015, 2023), Brand and Lang (2015) also point out that finance capital has discovered agriculture, land, infrastructure and environmental protection as new areas of investment, creating opportunities for a few and threatening the livelihoods of many, especially in the Global South.

1.2 Background Developments at the Global Level

For Dunlap and York (2008), the presumed lack of widespread concern for the environment in Less Developed Countries (LDCs) follows from the assumption in many current global values studies that environmental quality is a higher-order, quality-of-life value that poor people struggling to meet basic needs cannot afford to support. Dunlap and York (2008) point out that leading values research (see below) assumes a lack of environmental concern among citizens in the world's poorer nations. Postmaterialist, pro-environmental values, the argument goes, are far more prevalent in wealthy nations than in poor ones.

And yet, recent advances in environmental political economy, which examine the relationships between variables such as globalization, carbon emissions, population growth, urbanization and so on, would suggest that global values research would also be obliged to produce evidence focusing specifically on the growing importance of the ecology movement in the countries of the Global South.

For example, in a widely read study on the political ecology of the environment, Jorgenson (2012) argues that transnational corporations are building new or acquiring existing facilities in less developed countries—such as in Asia and Latin America—to take advantage of lower production costs and more permissive environmental laws. This shift in production, Jorgenson argues, has contributed to an increase in carbon emissions in less developed countries, even though much of the products are consumed in developed countries. The modest, relative decoupling between total emissions and development in countries of the Global North may in fact be linked to two sets of structural processes in countries of the Global South: (1) the intensification of the relationship between per capita emissions and economic development, and (2) the temporal stability of the strong relationship between total emissions and economic development. Therefore, Jorgenson (2012) argues that the dynamics of the global economic system require that a conceptualization of shifts be incorporated into the assessment of environment-economic development relationships.

The study of the interaction between level of development—or if you wish—existential security—and the state of the environment involves several highly intricate methodological issues. In a highly influential study on this subject, Liddle (2014) critically discusses the assumptions of the Environmental Kuznets Curve (EKC; see also https://www.sciencedirect.com/topics/economics-econometrics-and-finance/environmental-kuznets-curve). Dinda (2004) in his highly influential study on the subject, stipulated that the Environmental Kuznets Curve (EKC) hypothesis postulates an inverted U-shaped relationship between different pollutants and per capita income, i.e. environmental pressure increases up to a certain level as income goes up; after that, it decreases. Dinda (2004) highlights that the common point of all the studies is the assertion that the environmental quality deteriorates at the early stages of economic development/growth and subsequently improves at the later stages. In other words, environmental pressure increases faster than income at early stages of development and slows down relative to GDP growth at higher income levels. Dinda (2004) discusses the various explanations for this EKC (1) the progress of economic development, from clean agrarian economy to polluting industrial economy to clean service economy; (2) tendency of people with higher income having higher preference for environmental quality, etc.

Liddle (2014) highlights in this context that the empirical consideration of many cross-sectional units combined with observations taken at many time intervals (i.e. time series—cross-sectional or panel data) offers substantial advantages over simple cross-sectional analysis, e.g. (1) using country and/or time-fixed effects to control for some omitted variables (i.e. factors that may affect emissions that are not captured by variables specified in the regression model, such as economic shocks, changes in population policies, or other country-specific development pathways); (2) increasing substantially the degrees of freedom; and (3) allowing for dynamic modelling (e.g. estimating short-run and long-run effects). However, as Liddle (2014) also underlines, employing such time series and cross-sectional data both introduces statistical challenges and provides opportunities to address those challenges and other modelling issues: namely, serial correlation, non-stationarity, cross-sectional dependence, heterogeneity, nonlinearities, and endogeneity. Liddle (2014) also makes the point that a variable, which is particularly strongly related to global value change, population growth (see also Tausch et al., 2014) and its impact on the level of national carbon emissions has not been explicitly explored. Liddle (2014) also correctly makes the point that urbanization may lead to higher emissions/energy consumption through urbanization's association with industrialization—i.e. the shift from agriculture to industry and services. The co-evolving movement of people from rural to urban areas and from agricultural to industrial employment causes energy consumption to increase in three ways: (1) agricultural operations must mechanize as they become less labour intensive; (2) urbanization spatially separates food consumers from food producers, thus necessitating a transport requirement that did not exist under traditional agriculture and settlement patterns; and (3) modern industry/manufacturing uses more energy per unit of output and per worker than does traditional agricultural and manufacturing.

Such shifts will even increase the importance of the question, how economic changes in the Global South affect the strength of the ecology movement in the Global South.

In a recent study on the globalization-oriented drivers of the global environmental crisis, Tausch and Heshmati (2013) have closed the circuit from the usual studies on globalization and the usual studies on environmental degradation. The study showed severe detrimental effects of "globalization" measurement yardsticks, like high foreign savings, free production zones employment as % of total population, Multinational Corporation (MNC) penetration, and large-scale immigration on environmental data.

1.3 The World Values Survey Analyses on Environmental Movements

In the following, we will first attempt to provide our readers with a brief synthesis of established wisdom on global opinion surveys of environmental concerns (Abramson & Inglehart, 2009; Dunlap, 2008; Dutcher, 2007; Franzen & Vogl, 2013; Inglehart, 2009; Inglehart & Baker, 2000). Inglehart (see Abramson & Inglehart, 2009; Inglehart, 2009; Inglehart & Baker, 2000) maintains the position that value priorities in advanced industrial society will tend to shift away from materialist concerns about economic and physical security toward greater emphasis on freedom, self-expression, and the quality of life, or postmaterialist values. Inglehart (see Abramson & Inglehart, 2009; Inglehart 2009; Inglehart & Baker, 2000) argues that differences in the formative socialisation of young Europeans and their elders have led younger birth cohorts to place a relatively high priority on freedom and self-expression, and suggests that future intergenerational population replacement would bring a shift towards new value priorities. The growth of postmaterialist values will contribute to a decline of social class voting and to the rise of new social movements, particularly environmentalist movements and parties. Changing value priorities may, Inglehart maintains (see Abramson & Inglehart, 2009; Inglehart, 2009; Inglehart & Baker, 2000) be reshaping the nature of political cleavages and the political meaning of left and right, giving rise to a new politics axis. This new axis, according to Inglehart (see Abramson & Inglehart, 2009; Inglehart, 2009; Inglehart & Baker, 2000), cuts across the traditional left-right dimension, characterized by radical reform parties and movements at one pole and right authoritarian parties and movements like the Christian Coalition, the National Front, and the *Republikaner* at the other. Inglehart (see Abramson & Inglehart, 2009; Inglehart, 2009; Inglehart & Baker, 2000) demonstrates that there has been a clear trend toward postmaterialism largely resulting from intergenerational population replacement. Moreover, the growth of postmaterialism has occurred despite, rather than because of, rising levels of unemployment.

For Inglehart (see Abramson & Inglehart, 2009; Inglehart, 2009; Inglehart & Baker, 2000), the shift from materialist to postmaterialist values is not a uniquely Western phenomenon. Rather, it is found in societies with widely different institutions and cultural traditions. The rise of postmaterialist values is closely linked with prosperity and seems to occur wherever a society has experienced enough economic growth in recent decades so that the younger birth cohorts have experienced significantly greater economic security during their formative years than older cohorts experienced. Intergenerational value differences reflect a society's rate of economic growth. Economic growth, of course, is only one factor contributing to security or insecurity, but it happens to be (1) an important part of the story and (2) one for which we have relatively good cross-national and cross-time data. Inglehart (see Abramson & Inglehart, 2009; Inglehart, 2009; Inglehart & Baker, 2000) thinks that war, domestic upheaval, and ethnic conflict can also have a major impact on feelings of security, but precisely because they tend to be situation-specific (and are less readily quantified), they are more difficult to analyse empirically. Inter-generational differences are remarkably robust. According to Inglehart (see Abramson & Inglehart, 2009; Inglehart, 2009; Inglehart & Baker, 2000), in Western Europe, clear and substantial differences between the values of younger and older birth cohorts persisted through the recessions of the mid-1970s and the early 1980s. The postmaterialist value shift does not simply reflect current conditions: it also has a long-term component that seems to reflect the distinctive formative circumstances that given birth cohorts experienced as much as 40 or 50 years ago.

Based on their data analysis of the World Values Survey and Gallup's 24-nation "Health of the Planet" (HOP) survey, for Dunlap and York (2008) the crucial issue is that both conventional wisdom and social science explanations of environmental concern as stemming from postmaterialist values which would predict consistently positive relationships between citizen concern for the environment and levels of national affluence, but clearly the first three waves of the now seven waves of the World Values Survey (WVS; see methodology section, below) do not produce supportive evidence for either. When one considers that many of the WVS items seem biased in favour of yielding more pro-environmental responses among the publics of wealthy than poor nations, the results become even more noteworthy. Given the emphasis that both spokespersons of postmaterialism, such as Inglehart, place on public willingness to pay for environmental protection, and that the most straightforward indicators of such willingness are consistently (if not always significantly) negatively correlated with national affluence, for Dunlap and York (2008) the WVS results are particularly damaging—and even puzzling.

Dunlap and York (2008) conclude that those who have followed the rapidly accumulating evidence of citizen action on behalf of environmental protection in poor and developing nations around the world will not be surprised that environmental activism in these countries is often reflective of widespread public sentiment. Clearly both environmental activism and public support for environmental protection have become global phenomena and are no longer—if they ever were—limited to the wealthy nations of the world.

Dunlap and York (2008) also maintain that while it may take different shapes, concern for the environment has obviously spread well beyond wealthy nations, and it is time for both policymakers and social scientists to revise their views accordingly. To conceptualize environmental quality as something that only the wealthy can afford, and that is of little concern to the poor, does violence to the facts.

Franzen and Vogl (2013) analysed the development of environmental concern by using the three waves of the environmental modules of the International Social Survey Programme. The results show that environmental concern is closely correlated with the wealth of the nations. However, environmental concern decreased in almost all nations slightly during the last two decades. The decline was lower in countries with improving economic conditions suggesting that economic growth helps to maintain higher levels of environmental concern. The Franzen and Vogl (2013) results show that GDP has a positive effect on respondents' environmental concern and reconfirms the finding obtained with the cross-sectional data. Overall, environmental concern decreased slightly in almost all countries (the exception is Chile). However, the decrease was weaker in countries that experienced larger increases in GDP since 1993. This finding, Franzen and Vogl (2013) argue, is compatible with results obtained from time series analysis of public attitudes towards climate change in the United States. The controversy of political elites, particularly the scepticism regarding climate change by Republican leaders, contributed most strongly to the decrease. Franzen and Vogl (2013) think it is very much likely that after the 2008, the financial crisis pulled attention away from environmental concerns.

The fact that environmental concern decreased during the last two decades is, Franzen and Vogl (2013) maintain, of course bad news for the prospect of conserving the planet. It suggests that governments willing to implement measures for environmental protection will find it increasingly difficult to receive public support.

1.4 Methodology and Data

1.4.1 The World Values Survey and European Values Survey Data

Launched in 1981, the World Values Survey (WVS) is a series of nationally representative surveys conducted in nearly 100 countries, covering almost 90% of the world's population, using a common questionnaire on the attitudes of the world's population towards religion, politics, economics, society, education, prejudice, gender and sexuality, and the family. The WVS is the largest non-commercial, cross-national, time-series survey of human beliefs and values ever conducted, and currently includes interviews with nearly 400,000 respondents (Inglehart, 2020).

According to the current documentation of the WVS (https://www.worldvaluessurvey.org/WVSDocumentationWV7.jsp) and our own macro-quantitative data (see below), the WVS currently captures the opinions of more than 5 billion global residents, or about 66% of the world's population.

The current study uses the well-established methodology of analysing data from international surveys, again in the World Values Survey, as already presented in detail in the study by Tausch et al. (2014). We would like to emphasize that, in addition to comparing percentages and means in cross-tabulations, the present study makes particular use of the method of partial correlations and promax factor analysis. As can be seen in Tausch et al. (2014), promax factor analysis is particularly suitable for extracting dimensions of variables that may be correlated with each other from a dataset with many variables. Table 1.1 of our chapter shows the date of the WVS samples as well as the sample size *N* (see Table 1.1 and Fig. 1.1).

The identical version of the joint EVS/WVS is accessible through two data service points:

- EVS/GESIS: via the GESIS Data Collection at GESIS—Leibniz Institute for the Social Sciences (data download page)
- WVSA: via the WVS website

1.4.2 Methodology

Our research attempt is of course guided by the vast traditions of mathematical-statistical analysis in opinion survey research (see Tausch et al., 2014, see furthermore Abdi, 2003; Basilevsky, 2009; Braithwaite & Law, 1985; Brenner, 2016; Browne, 2001; Fabrigar et al., 1999; Hedges & Olkin, 2014; Kline, 2014; Knippenberg, 2015; McDonald, 2014; Mulaik, 2009; Suhr, 2012; Yeşilada & Noordijk, 2010).

Our methodological approach is within a more general framework to study global values with the methodology of comparative and opinion-survey-based political science (Brenner, 2016; Inglehart, 2018, 2020; Knippenberg, 2015). Our methodology of evaluating the opinions of global publics from global surveys is in addition based on recent advances in mathematical statistical factor analysis (Basilevsky, 2009; Hedges & Olkin, 2014; Kline, 2014; McDonald, 2014; Mulaik, 2009). Such studies allow to project the underlying structures of the relationships between the variables.

Current methodology of the social sciences makes it clear that besides factor analysis, there are also other powerful tools of multivariate analysis available to test complex relationships between an independent variable and independent variables (Tabachnick & Fidell, 2001). In our case, we used partial correlation analysis and factor analysis. Throughout this work, we use standard partial regression analysis to

Table 1.1 Our surveys from the World Values Survey and the European Values Study

Country	Year of survey	$N =$
Albania	2018	1435
Andorra	2018	1004
Argentina	2017	1003
Armenia	2018, 2021	2723
Australia	2018	1813
Austria	2018	1644
Azerbaijan	2018	1800
Bangladesh	2018	1200
Belarus	2018	1548
Bolivia	2017	2067
Bosnia and Herzegovina	2019	1724
Brazil	2018	1762
Bulgaria	2017	1558
Canada	2020	4018
Chile	2018	1000
China	2018	3036
Colombia	2018	1520
Croatia	2017	1487
Cyprus	2019	1000
Czechia	2017	1811
Denmark	2017	3362
Ecuador	2018	1200
Egypt	2018	1200
Estonia	2018	1304
Ethiopia	2020	1230
Finland	2017	1199
France	2018	1870
Georgia	2018	2194
Germany	2017, 2018	3698
Great Britain	2018	1788
Greece	2017	1200
Guatemala	2020	1229
Hong Kong SAR	2018	2075
Hungary	2018	1514
Iceland	2017	1624
Indonesia	2018	3200
Iran	2020	1499
Iraq	2018	1200
Italy	2018	2277
Japan	2019	1353
Jordan	2018	1203
Kazakhstan	2018	1276

(continued)

Table 1.1 (continued)

Country	Year of survey	$N =$
Kenya	2021	1266
Kyrgyzstan	2020	1200
Latvia	2021	1335
Lebanon	2018	1200
Libya	2022	1196
Lithuania	2018	1448
Macau SAR	2019	1023
Malaysia	2018	1313
Maldives	2021	1039
Mexico	2018	1741
Mongolia	2020	1638
Montenegro	2019	1003
Morocco	2021	1200
Myanmar	2020	1200
Netherlands	2017, 2022	4549
New Zealand	2020	1057
Nicaragua	2020	1200
Nigeria	2018	1237
North Macedonia	2019	1117
Norway	2018	1122
Pakistan	2018	1995
Peru	2018	1400
Philippines	2019	1200
Poland	2017	1352
Portugal	2020	1215
Puerto Rico	2018	1127
Romania	2018	2870
Russia	2017	3635
Serbia	2017, 2018	2545
Singapore	2020	2012
Slovakia	2017	1432
Slovenia	2017	1075
South Korea	2018	1245
Spain	2017	1209
Sweden	2017	1194
Switzerland	2017	3174
Taiwan ROC	2019	1223
Tajikistan	2020	1200
Thailand	2018	1500
Tunisia	2019	1208
Türkiye	2018	2415
Ukraine	2020	2901

(continued)

Table 1.1 (continued)

Country	Year of survey	$N =$
United States	2017	2596
Venezuela	2021	1190
Vietnam	2020	1200
Zimbabwe	2020	1215
Total $N =$	2017–2022	147,260

come to terms with the often-underestimated omitted variable bias in quantitative social science, always cross-checking significant correlations with already existing literature on the subject. Omitted variable bias indeed is a serious problem in the discipline (see Tausch et al., 2014).

Our research also uses myriads of cross-national development data and was written in the well-established tradition of macro-quantitative and time-series development research (see below). We especially mention what can be termed "development accounting" (see: Tausch & Heshmati, 2013). We project our research results also at the level of choropleth maps.[1]

In general, our entire work firmly shares the established methodology of global values and comparative opinion research (Tausch et al., 2014).

Our main statistical calculations relied on simple cross tables, comparisons of means, bi-variate and partial correlation analyses, factor analysis (oblique factor rotations based on promax factor analysis) (Abdi, 2003; Babones, 2014; Basilevsky, 2009; Browne, 2001; Clauß & Ebner, 1970; Fabrigar et al., 1999; Hedges & Olkin, 2014; Kline, 2014; Suhr, 2012; Tabachnick & Fidell, 2001; for a condensed survey, see also Tausch et al., 2014).

For the algorithm of partial correlation analysis and promax factor analysis, we refer our readers to IBM-SPSS; Hendrickson and White (1964); and Morrison (1976).

1.4.3 *Parametric Indicators*

But our destabilization indicator is a so-called "parametric indicator", which—in every day plain language—combines the data with the aim of multivariate statistical analysis (see Tausch et al., 2014). Such a parametric indicator relies on advanced statistical methods, such as principal components analysis (see, again, Tausch et al., 2014). Such an analysis extracts an overriding indicator, mathematically best

[1] The choropleth maps of this chapter were drawn using the free software developed by Robert Mundigl, available at https://www.clearlyandsimply.com/

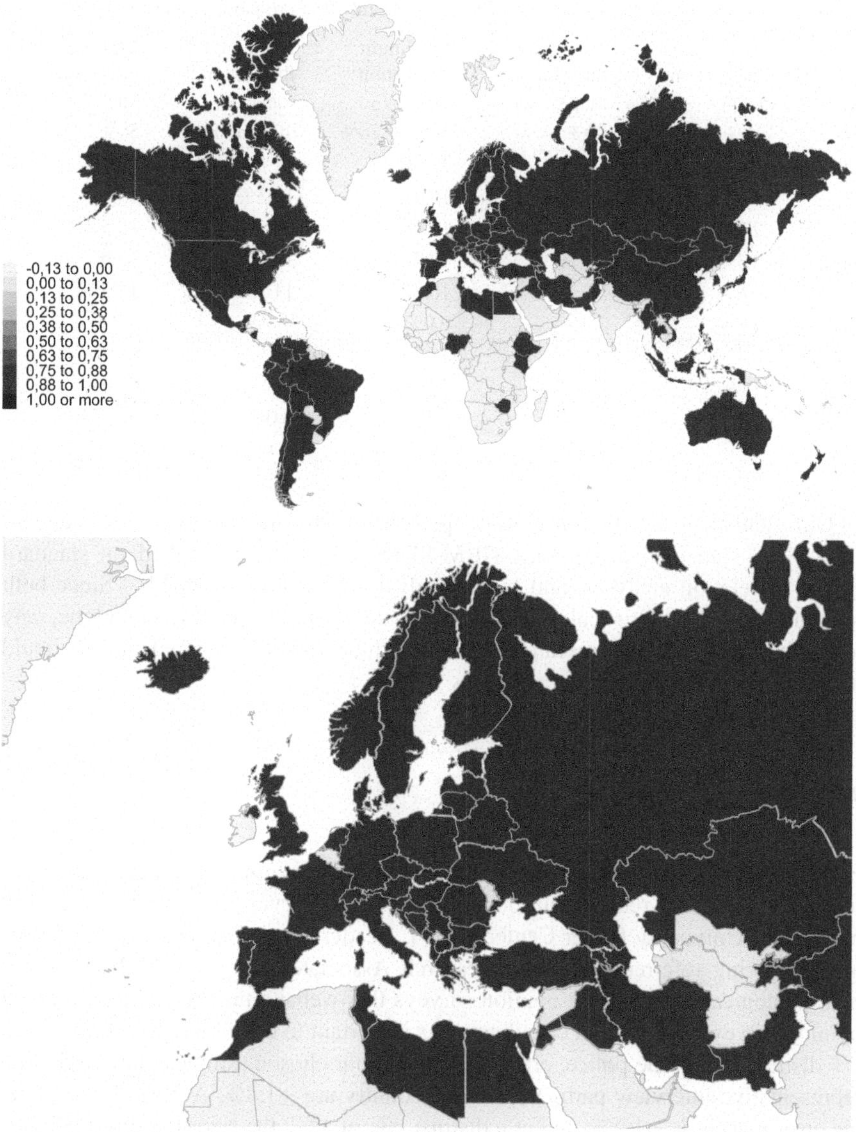

Fig. 1.1 The global reach of our multivariate models, based on World Values Survey Wave 7 and the European Values Study (WVS Cross-National Wave 7 spss v4 0.zip)

representing the component variables and their correlation matrix. Our parametric index thus relies on the original survey respondents of the survey, and calculates the country results, based on principal components factor scores.

Table 1.2 Maximum ranges of variation for survey results (the probability of error is 5%)

Sample size	Maximum fluctuation ranges (±)	Maximum fluctuation ranges (±)	Maximum fluctuation ranges (±)	Maximum fluctuation ranges (±)	Maximum fluctuation ranges (±)
N	10% or 90%	20% or 80%	30% or 70%	40% or 60%	50%
20	13.1%	17.5%	20.1%	21.5%	21.9%
30	10.7%	14.3%	16.4%	17,5%	17.9%
40	9.3%	12.4%	14.2%	15.2%	15.5%
50	8.3%	11.1%	12.7%	13.6%	13.9%
75	6.8%	9.1%	10.4%	11.1%	11.3%
100	5.9%	7.8%	9.0%	9.6%	9.8%
250	3.7%	5.0%	5.7%	6.1%	6.2%
500	2.6%	3.5%	4.0%	4.3%	4.4%
1000	1.9%	2.5%	2.8%	3.0%	3.1%
2000	1.3%	1.8%	2.0%	2.1%	2.2%

Our statistical calculations were performed by the routine and standard IBM-SPSS statistical programme (IBM-SPSS XXVIII),[2] and relied on standard partial correlation analyses, and factor analysis (Tausch et al., 2014).[3] Since both our data and the statistical methods used are available around the globe, any researcher can repeat our research exercise with the available open data and should be able to reproduce the same results as we did.

1.4.4 Error Margins

For the calculation of error margins of the representative opinion survey, readers are referred to the easily readable introduction to opinion survey error margins, prepared by Cornell University Roper Center (2017). Readers more interested in the details are also being referred to Langer Research Associates n.d.[4] On the basis of the methodological literature on opinion surveys this website makes available a direct opinion survey error margin calculator. It is important to recall that, for example at a 5% distrust rate in the police, error margins for our chosen samples of around 1000 representative interview partners for each country are ±1.4%. A 10% distrust rate, the error margin is ±1.9%; and at a distrust rate of 15% the error margin is ±2.2%; see Langer Research Associates n.d. That error margins differ according to reported rates of, say, distrust in the police, is an important fact of opinion survey research theory, often forgotten to be mentioned in the public debate.

[2]https://www-01.ibm.com/software/at/analytics/spss/

[3]https://www.ibm.com/analytics/spss-statistics-software

[4]https://www.langerresearch.com/moe/

Keeping in line with standard traditions of empirical opinion survey research (Tausch et al., 2014), for all analysed groups and sub-groups, a minimum sample size of at least 30 respondents per country had to be available to be able to attempt reasonable predictions (Clauß & Ebner, 1970) (See also Table 1.2).

1.4.5 Dimensions and Variables from the World Values Survey and European Values Survey

We used the following World Values Survey Data:

- Member: Belong to conservation, the environment, ecology, animal rights
- No confidence: The Environmental Protection Movement
- Democracy: Governments tax the rich and subsidize the poor.
- Democracy: Religious authorities interpret the laws.
- Democracy: People choose their leaders in free elections.
- Democracy: People receive state aid for unemployment.
- Democracy: Civil rights protect people's liberty against oppression.
- Democracy: Women have the same rights as men.
- Democracy: The state makes people's incomes equal
- Democracy: People obey their rulers
- Gender
- Year of birth
- Highest educational level attained—Respondent: ISCED-code one digit

1.4.6 Cross-National Data

In the explanation of the partial correlations of environmentalism, due emphasis was given to dependency and world system approaches to development, which received a large-scale empirical confirmation (see also Tausch & Heshmati, 2013). Earlier, well-known datasets for these investigations were

- Ballmer-Cao & Scheidegger, 1979
- Müller & Bornschier, 1988
- Tausch, 2012 [*International Macroquantitative Data*. Faculty of Economics, Corvinus University of Budapest (with data definitions and sources). Free download available from http://www.uni-corvinus.hu/index.php?id=47854&tx_ttnews%5Btt_news%5D=0&tx_ttnews%5BbackPid%5D=31638&tx_ttnews%5BcalendarYear%5D=2012&tx_ttnews%5BcalendarMonth%5D=6&cHash=af8ef6888f7c9922b83b113f71c1ca32]

More recent databases, integrating dependency and world system approaches with conventional economic theories and the sociology of world values research are:

- Tausch, 2019. [*Migration from the Muslim world to the West: Its most recent trends and effects*. Jewish Political Studies Review, 30(1–2), 65–225, available at http://jcpa.org/article/migration-from-the-muslim-world-to-the-west-its-most-recent-trends-and-effects/ (with data definitions and sources). Free data download available from https://www.academia.edu/37568941/Migration_from_the_Muslim_World_to_the_West_Its_Most_Recent_Trends_and_Effects]
- Tausch & Heshmati, 2013 [*Globalization, the human condition, and sustainable development in the twenty-first century: Cross-national perspectives and European implications*. London, New York, and Delhi: Anthem Press (with data definitions and sources). Free data download available from https://www.academia.edu/35044095/Globalization_the_human_condition_and_sustainable_development_in_the_21st_Century._Cross-national_perspectives_and_European_implications_Codebook_and_EXCEL_data_file]

The present data set, which we used in our analysis, is available in EXCEL format data (https://www.researchgate.net/publication/368810044_RESILIENCE_2023_EXCEL_DATA_PUBLIC_ACCESS_for_the_publication_Destabilizing_forces_a_MENA_country_regional_survey_and_a_global_comparison).

1.5 Results from the World Values Survey and European Values Survey

Table 1.3 and Fig. 1.2 clearly show the rising tide of environmental movements in the Global South, squarely contradicting many of the prognosis of earlier World Values Survey research.

Table 1.4 documents the high confidence in the environmental movement in many countries and territories of the Global South, especially in East and Southeast Asia. In general terms, trust in the environmental movement is lowest in some Arab countries, in Southeastern and Eastern Europe, and in some countries of Latin America and the Carribean.

In the following analysis, we attempted another factor analysis focusing on

- Member: Belong to conservation, the environment, ecology, animal rights
- No confidence: The Environmental Protection Movement
- Democracy: Governments tax the rich and subsidize the poor.
- Democracy: Religious authorities interpret the laws.
- Democracy: People choose their leaders in free elections.
- Democracy: People receive state aid for unemployment.
- Democracy: Civil rights protect people's liberty against oppression.
- Democracy: Women have the same rights as men.
- Democracy: The state makes people's incomes equal
- Democracy: People obey their rulers
- Gender

Table 1.3 Member: belong to conservation, the environment, ecology, animal rights (0 = 0%; 1.0 = 100%)

	Member: Belong to conservation, the environment, ecology, animal rights	*N*
Kenya	0.394	1203
Indonesia	0.349	3188
Colombia	0.346	1520
Thailand	0.320	1500
Guatemala	0.263	1192
Malaysia	0.263	1313
Nicaragua	0.256	1200
Libya	0.234	1170
Ethiopia	0.220	1216
Mongolia	0.218	1638
Morocco	0.218	1200
Tajikistan	0.216	1200
Nigeria	0.210	1211
Switzerland	0.206	3142
United States	0.196	2573
Zimbabwe	0.191	1205
New Zealand	0.181	982
Philippines	0.174	1199
Pakistan	0.173	1796
Netherlands	0.172	4385
Hong Kong SAR	0.166	2075
Puerto Rico	0.165	1127
Taiwan ROC	0.164	1223
Australia	0.157	1781
Mexico	0.155	1738
Chile	0.154	1000
Bolivia	0.152	2045
Canada	0.150	4018
Cyprus	0.146	984
Denmark	0.144	3349
Sweden	0.142	1177
Slovenia	0.138	1071
Iran	0.128	1497
Czechia	0.125	1554
Ecuador	0.122	1186
Iceland	0.122	1616
North Macedonia	0.115	1092
Macau SAR	0.103	1017
Iraq	0.102	1164
Germany	0.100	3635
Great Britain	0.086	1787

(continued)

Table 1.3 (continued)

	Member: Belong to conservation, the environment, ecology, animal rights	*N*
Finland	0.079	1178
Argentina	0.070	1003
Norway	0.066	1122
Maldives	0.061	1012
Myanmar	0.060	1200
Croatia	0.058	1427
Tunisia	0.058	1174
Ukraine	0.058	2820
Serbia	0.053	2391
Bangladesh	0.051	1173
South Korea	0.051	1245
Turkey	0.051	2371
Romania	0.050	2741
Venezuela	0.050	1190
Peru	0.049	1389
Singapore	0.049	2004
Spain	0.049	1201
France	0.047	1868
Hungary	0.046	1512
Kazakhstan	0.043	1196
Lebanon	0.043	1200
Austria	0.040	1637
China	0.039	3023
Armenia	0.038	2716
Brazil	0.037	1719
Vietnam	0.037	1200
Andorra	0.036	1002
Greece	0.035	1199
Jordan	0.035	1160
Kyrgyzstan	0.032	1194
Bulgaria	0.031	1435
Bosnia and Herzegovina	0.028	1696
Latvia	0.028	1332
Montenegro	0.027	997
Italy	0.026	2249
Poland	0.022	1346
Georgia	0.018	2178
Albania	0.017	1435
Lithuania	0.016	1413
Estonia	0.014	1304
Japan	0.014	1336
Russia	0.014	3585

(continued)

Table 1.3 (continued)

	Member: Belong to conservation, the environment, ecology, animal rights	*N*
Slovakia	0.010	1411
Belarus	0.009	1548
Portugal	0.009	1188
Azerbaijan	0.005	1775
Egypt	0.005	1199

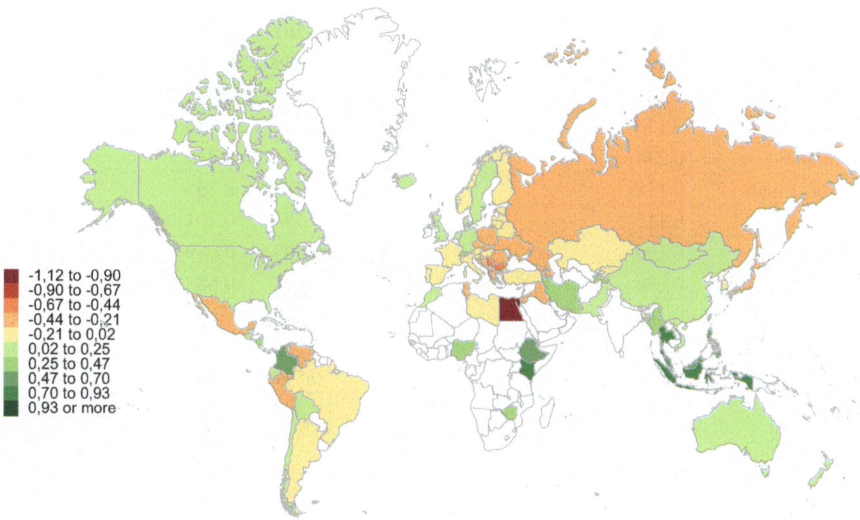

Fig. 1.2 Environmental protection movement

- Year of birth
- Highest educational level attained—Respondent: ISCED-code one digit

as a "microscopic close-up" of the reality of environmental movements today. All statistical quality indicators, including the Bartlett test of sphericity, confirm the good and reliable statistical quality of our analysis, which explains 58.822% of the total variance. For the sake of simplicity, we present only the structure matrix of the promax factor analysis, where all factors with an eigenvalue greater than 1.0 are reported in Table 1.5, and the country factor scores in Table 1.6. The results in Table 1.6 are indicated by the factor scores of factor 3, the environmental movement, with the civil societies of the following 20 countries at the top:

1. Indonesia
2. Thailand
3. Kenya

Table 1.4 Confidence in the environmental movement (0 = 0%; 1.0 = 100%)

Xx	Confidence: The Environmental Protection Movement	Xx	xx	Xx
Xx	A great deal	Quite a lot	A great deal + quite a lot	*N*
Vietnam	0.177	0.719	0.896	1162
China	0.152	0.663	0.815	3016
Myanmar	0.323	0.484	0.807	1200
Indonesia	0.281	0.516	0.797	3173
Philippines	0.309	0.480	0.789	1199
Malaysia	0.180	0.595	0.775	1310
Iran	0.263	0.501	0.764	1474
Macau SAR	0.106	0.636	0.742	1016
Singapore	0.107	0.615	0.722	1750
Taiwan ROC	0.126	0.589	0.715	1207
Ethiopia	0.234	0.481	0.715	1147
Great Britain	0.118	0.587	0.705	1734
Sweden	0.117	0.584	0.701	1160
Thailand	0.196	0.493	0.689	1127
Andorra	0.088	0.601	0.689	994
Puerto Rico	0.344	0.335	0.679	1113
Iceland	0.102	0.561	0.663	1566
Estonia	0.090	0.571	0.661	1115
Latvia	0.056	0.598	0.654	1250
Kyrgyzstan	0.144	0.508	0.652	1133
South Korea	0.057	0.591	0.648	1245
Bangladesh	0.172	0.476	0.648	1081
Zimbabwe	0.214	0.429	0.643	1175
Germany	0.086	0.556	0.642	3477
France	0.087	0.554	0.641	1779
Switzerland	0.072	0.567	0.639	3124
New Zealand	0.089	0.546	0.635	953
Ecuador	0.198	0.434	0.632	1190
Brazil	0.173	0.454	0.627	1575
Spain	0.131	0.494	0.625	1159
Kenya	0.272	0.349	0.621	1237
Canada	0.100	0.519	0.619	4018
Portugal	0.071	0.548	0.619	1107
Hong Kong SAR	0.085	0.533	0.618	2057
Denmark	0.083	0.534	0.617	3277
Austria	0.131	0.482	0.613	1596
Lithuania	0.037	0.576	0.613	1226
Belarus	0.110	0.502	0.612	1366
Kazakhstan	0.159	0.452	0.611	1193
Tajikistan	0.250	0.360	0.610	1048

(continued)

Table 1.4 (continued)

Xx	Confidence: The Environmental Protection Movement	Xx	xx	Xx
Xx	A great deal	Quite a lot	A great deal + quite a lot	*N*
Slovakia	0.061	0.548	0.609	1338
Maldives	0.191	0.414	0.605	1025
Pakistan	0.224	0.378	0.602	1734
Chile	0.178	0.423	0.601	971
Argentina	0.135	0.462	0.597	865
Italy	0.106	0.491	0.597	2169
Australia	0.095	0.501	0.596	1788
Norway	0.090	0.504	0.594	1103
Cyprus	0.109	0.479	0.588	890
Azerbaijan	0.097	0.487	0.584	1523
Nigeria	0.190	0.377	0.567	1211
Turkey	0.088	0.472	0.560	2285
Hungary	0.114	0.443	0.557	1419
Nicaragua	0.253	0.304	0.557	1200
United States	0.116	0.434	0.550	2565
Finland	0.055	0.491	0.546	1160
Japan	0.052	0.486	0.538	1109
Russia	0.112	0.413	0.525	3201
Netherlands	0.056	0.468	0.524	4118
Slovenia	0.069	0.441	0.510	1031
Georgia	0.113	0.386	0.499	2035
Poland	0.077	0.400	0.477	1207
Bolivia	0.164	0.302	0.466	1950
Greece	0.046	0.420	0.466	1134
Jordan	0.137	0.328	0.465	879
Colombia	0.226	0.234	0.460	1520
Venezuela	0.092	0.366	0.458	1190
Morocco	0.090	0.352	0.442	1200
Peru	0.100	0.336	0.436	1323
North Macedonia	0.085	0.349	0.434	1060
Ukraine	0.046	0.387	0.433	2520
Armenia	0.069	0.357	0.426	2496
Mongolia	0.096	0.319	0.415	1498
Bosnia and Herzegovina	0.075	0.326	0.401	1673
Montenegro	0.061	0.338	0.399	958
Romania	0.096	0.302	0.398	2562
Guatemala	0.093	0.281	0.374	1229
Mexico	0.126	0.245	0.371	1696

(continued)

Table 1.4 (continued)

Xx	Confidence: The Environmental Protection Movement	Xx	xx	Xx
Xx	A great deal	Quite a lot	A great deal + quite a lot	*N*
Tunisia	0.174	0.194	0.368	1003
Croatia	0.048	0.309	0.357	1426
Iraq	0.137	0.188	0.325	1053
Albania	0.062	0.260	0.322	1345
Serbia	0.059	0.262	0.321	2373
Czechia	0.048	0.266	0.314	1622
Libya	0.078	0.206	0.284	1054
Bulgaria	0.025	0.253	0.278	1359
Lebanon	0.063	0.159	0.222	1164
Egypt	0.009	0.140	0.149	681

4. Malaysia
5. Philippines
6. Colombia
7. Ethiopia
8. Puerto Rico
9. Nicaragua
10. Tajikistan
11. Iran
12. Zimbabwe
13. Nigeria
14. Switzerland
15. Myanmar
16. Taiwan ROC
17. New Zealand
18. Sweden
19. United States
20. Vietnam

Table 1.5 further highlights our research results.

1.6 Perspectives and General Conclusions

Our analysis has revealed the following dramatic trends:

Environmentalism is increasingly part of the political reality of the Global South, especially in East and Southeast Asia, as shown in Fig. 1.2.

Environmentalism thrives in an atmosphere of trust.

Table 1.5 Promax factor analysis of democracy, environmentalism, gender and generations

	Support of liberal democracy	Religious authoritarian current	Environmental protection movement	Elderly and high educated people	Female factor
Member: Belong to conservation, the environment, ecology, animal rights	−0.110	0.011	0.751	0.070	−0.155
No confidence: The Environmental Protection Movement	−0.110	−0.085	−0.740	−0.014	−0.175
Democracy: Governments tax the rich and subsidize the poor	0.459	0.477	0.025	−0.124	0.036
Democracy: Religious authorities interpret the laws	−0.212	0.710	0.038	0.009	0.049
Democracy: People choose their leaders in free elections	0.765	−0.024	−0.024	−0.003	−0.042
Democracy: People receive state aid for unemployment	0.650	0.285	−0.042	−0.142	0.099
Democracy: Civil rights protect people's liberty against oppression	0.754	0.089	0.025	0.028	−0.016
Democracy: Women have the same rights as men	0.740	−0.027	0.011	−0.016	0.058
Democracy: The state makes people's incomes equal	0.369	0.642	0.000	−0.095	0.154
Democracy: People obey their rulers	0.108	0.668	0.068	−0.019	−0.043
Sex	−0.003	0.016	0.007	−0.004	0.961
Year of birth	−0.151	0.193	0.071	0.800	0.052
Highest educational level attained—Respondent: ISCED-code one digit	0.192	−0.366	0.008	0.661	−0.097

The Arab world, the Orthodox world and the orbit of the European Union are (no longer) at the centre of the expansion of the environmental movement.

Some partial correlations (with the Human Development Index and its squared constant) with our macro-societal data support these claims, as presented in Table 1.7.

Table 1.6 Factor scores of democracy, environmentalism, gender and generations

Country (ISO 3166-1 Numeric code)	Support of liberal democracy	Religious authoritarian current	Environmental protection movement	Elderly and high educated people	Female factor
Indonesia	−0.374	0.995	0.930	−0.131	−0.013
Thailand	−1.074	−0.113	0.824	−0.489	−0.064
Kenya	−0.848	0.098	0.778	0.709	−0.165
Malaysia	−1.473	0.370	0.585	0.162	0.009
Philippines	−0.568	0.672	0.505	−0.338	0.103
Colombia	−0.930	−0.247	0.481	0.142	−0.168
Ethiopia	0.158	0.888	0.476	0.208	−0.051
Puerto Rico	0.250	−0.283	0.384	−0.002	0.134
Nicaragua	−0.655	0.139	0.379	0.216	−0.010
Tajikistan	0.111	0.726	0.378	0.389	−0.067
Iran	0.110	0.661	0.317	0.397	0.040
Zimbabwe	−0.064	0.701	0.314	−0.021	−0.026
Nigeria	−0.389	0.609	0.300	0.203	−0.149
Switzerland	0.473	−0.512	0.276	−0.085	−0.079
Myanmar	0.080	1.005	0.269	−0.195	0.164
Taiwan ROC	0.219	−0.249	0.265	−0.026	−0.041
New Zealand	0.279	−0.884	0.244	−0.169	−0.112
Sweden	0.493	−0.677	0.231	−0.131	−0.172
United States	−0.037	−0.583	0.210	0.440	−0.316
Vietnam	0.060	0.872	0.185	0.135	0.266
Pakistan	−0.132	1.291	0.179	−0.020	−0.116
Ecuador	−0.730	0.025	0.169	0.124	0.000
Denmark	0.511	−0.515	0.169	−0.128	−0.084
Macau SAR	−0.026	−0.287	0.155	0.258	0.147
Chile	−0.309	0.112	0.143	0.013	−0.086
Iceland	0.679	−0.197	0.141	−0.093	−0.061
Australia	0.202	−0.796	0.139	−0.078	−0.020
Great Britain	0.156	−0.421	0.128	−0.230	0.007
Canada	0.288	−0.341	0.124	0.338	−0.149
China	0.302	0.237	0.122	−0.298	0.258
Hong Kong SAR	−0.087	−0.437	0.103	−0.350	0.006
Netherlands	0.351	−0.747	0.099	0.054	−0.173
Bolivia	−0.337	0.291	0.085	0.240	−0.115
Guatemala	−0.792	−0.109	0.060	0.745	−0.145
Singapore	−0.112	−0.306	0.057	0.173	0.011
Germany	0.709	−0.787	0.043	−0.241	−0.031
Mongolia	−0.726	−0.080	0.038	0.647	−0.099
Cyprus	0.217	0.104	0.035	0.327	−0.102
Morocco	−0.155	0.808	0.033	0.037	0.013
Bangladesh	0.223	1.545	0.008	−0.160	0.107

(continued)

Table 1.6 (continued)

Country (ISO 3166-1 Numeric code)	Support of liberal democracy	Religious authoritarian current	Environmental protection movement	Elderly and high educated people	Female factor
Slovenia	0.234	−0.560	−0.004	−0.224	0.025
Maldives	−0.419	0.577	−0.008	0.170	0.057
Argentina	0.085	0.475	−0.039	−0.217	0.029
Kazakhstan	−0.191	0.484	−0.046	0.404	0.116
Andorra	0.558	−0.105	−0.057	−0.108	−0.021
France	−0.079	−0.278	−0.058	−0.300	0.020
Brazil	−0.103	−0.204	−0.069	−0.102	0.091
South Korea	−0.159	−0.058	−0.071	0.000	0.037
Finland	0.487	−0.238	−0.076	−0.187	−0.040
Norway	0.552	−0.387	−0.076	−0.015	−0.018
Kyrgyzstan	−0.160	0.372	−0.091	0.465	0.213
Austria	0.585	−0.524	−0.095	−0.383	0.176
Spain	0.215	0.189	−0.095	−0.452	0.116
Libya	−0.414	0.686	−0.117	0.573	−0.271
Estonia	0.319	−0.427	−0.123	−0.196	0.152
Latvia	−0.022	−0.331	−0.129	−0.009	0.254
Hungary	0.125	−0.159	−0.162	−0.206	0.057
Italy	0.383	0.134	−0.163	−0.484	0.051
Lithuania	0.297	−0.081	−0.185	−0.108	0.236
Portugal	0.121	0.093	−0.193	−0.681	0.193
Belarus	0.165	−0.053	−0.194	0.279	0.065
Turkey	−0.272	0.338	−0.194	−0.176	0.095
North Macedonia	0.174	−0.277	−0.217	0.151	−0.055
Mexico	−0.832	−0.037	−0.236	−0.102	−0.138
Slovakia	−0.258	−0.100	−0.238	−0.340	0.230
Azerbaijan	0.234	0.015	−0.266	0.047	0.046
Japan	0.250	−0.928	−0.266	−0.220	−0.120
Czechia	−0.255	−0.760	−0.292	−0.171	−0.066
Peru	−0.355	0.164	−0.293	0.180	−0.098
Jordan	−0.366	0.700	−0.304	−0.054	−0.129
Russia	0.169	0.294	−0.311	0.260	0.090
Georgia	−0.148	−0.250	−0.319	−0.019	0.135
Tunisia	−0.332	0.395	−0.320	−0.223	0.011
Venezuela	−0.542	0.097	−0.322	0.252	0.005
Poland	0.223	−0.538	−0.331	−0.095	−0.070
Ukraine	0.358	0.108	−0.336	0.332	0.136
Greece	0.505	−0.105	−0.342	−0.443	−0.001
Romania	0.113	0.077	−0.367	−0.227	0.021
Armenia	−0.080	0.000	−0.397	0.211	0.055
Croatia	0.265	−0.314	−0.412	−0.211	0.091

(continued)

Table 1.6 (continued)

Country (ISO 3166-1 Numeric code)	Support of liberal democracy	Religious authoritarian current	Environmental protection movement	Elderly and high educated people	Female factor
Montenegro	−0.432	0.159	−0.415	−0.132	−0.028
Bosnia and Herzegovina	0.203	0.040	−0.428	−0.222	0.144
Iraq	−0.173	0.764	−0.434	0.085	−0.100
Serbia	−0.022	−0.374	−0.494	0.112	−0.032
Bulgaria	0.341	−0.645	−0.538	−0.316	−0.014
Lebanon	−0.350	0.145	−0.620	0.126	−0.079
Albania	0.507	−0.316	−0.628	−0.294	0.230
Egypt	−0.232	0.570	−0.895	0.189	−0.322

Table 1.7 Partial correlations with the Factor Scores: "Environmental protection movement"

HDI 2018 & HDI (2018)^2	Partial Corr Environmental Movement	Error p	df
Gallup poll about satisfaction: Education quality	0.809	<0.001	77
Gallup poll about satisfaction: Health care quality	0.781	<0.001	77
Gallup poll about satisfaction: Standard of living	0.759	<0.001	77
Gallup poll about satisfaction: Efforts to deal with the poor	0.689	<0.001	77
Gallup poll about satisfaction: Job	0.631	<0.001	77
Gallup poll about satisfaction: Overall life satisfaction index	0.624	<0.001	77
Gallup poll about satisfaction: Local labour market	0.603	<0.001	77
Gallup poll about satisfaction: Freedom of choice	0.568	<0.001	77
Distance to Belgium	0.514	<0.001	76
Gallup poll about satisfaction: Trust in national government	0.400	<0.001	74
Gender empowerment index value	0.399	0.003	51
Happy Planet Index, HPI	0.356	0.002	74
Share of Buddhists per total population	0.313	0.004	79
Share of Hindus per total population	0.308	0.005	79
DYN Human Development Index HDI 2010–2018	0.262	0.017	81
MENA countries	−0.240	0.029	81
EEA + EFTA	−0.308	0.006	76
Arab League	−0.324	0.003	81
Share of Orthodox Christians per total population	−0.355	0.001	79
Unemployment rate	−0.375	0.001	71
Total unemployment rate of immigrants (both sexes)	−0.522	<0.001	76
Absolute latitude	−0.584	<0.001	68

References

Abdi, H. (2003). *Factor rotations in factor analyses. Encyclopedia for research methods for the social sciences* (pp. 792–795).

Abramson, P., & Inglehart, R. F. (2009). *Value change in global perspective*. University of Michigan Press.

Babones, S. (2014). *Methods for quantitative macro-comparative research*. Sage.

Ballmer-Cao, T. H., & Scheidegger, J. (1979). *Compendium of data for world-system analysis: A sourcebook of data based on the study of MNCs, economic policy and national development*. Sociological Institute, University of Zurich.

Basilevsky, A. T. (2009). *Statistical factor analysis and related methods: Theory and applications* (Vol. 418). John Wiley & Sons.

Braithwaite, V. A., & Law, H. G. (1985). Structure of human values: Testing the adequacy of the Rokeach Value Survey. *Journal of Personality and Social Psychology, 49*(1), 250.

Brand, U. (2015). Brave green world: The green economy myths. Luxemburg Argumente, no. 3. https://www.rosalux.de/publication/38457/beautiful-green-world.html

Brand, U. (2023). *Structural limits to a green economy*. International Sociological Association, & International Sociological Association Forum. Available at https://futureswewant.net/ulrich-brand-green-economy/

Brand, U, & Lang, M (2015). Entry "green economy". In: Pattberg, Philipp/Zelli, Fariborz (eds.): Encyclopedia of global environmental politics and governance. : Edward Elgar.

Brenner, P. S. (2016). Cross-national trends in religious service attendance. *Public Opinion Quarterly, 80*, 563.

Browne, M. W. (2001). An overview of analytic rotation in exploratory factor analysis. *Multivariate Behavioral Research, 36*(1), 111–150.

Clauß, G., & Ebner, H. (1970). *Grundlagen der Statistik für Psychologen*. Volk und Wissen Volkseigener Verlag.

Dinda, S. (2004). Environmental Kuznets curve hypothesis: a survey. *Ecological Economics, 49*(4), 431–455.

Dunlap, R. (2008). The globalization of environmental concern and the limits of the postmaterialist values explanation: Evidence from four multinational surveys. *The Sociological Quarterly Tsq; Official Journal of the Midwest Sociological Society, 49*(3), 529–563.

Dunlap, R. E., & York, R. (2008). The globalization of environmental concern and the limits of the postmaterialist values explanation: Evidence from four multinational surveys. *The Sociological Quarterly, 49*(3), 529–563.

Dutcher, D. (2007). Connectivity with nature as a measure of environmental values. *Environment and Behavior, 39*(4), 474–493.

Fabrigar, L. R., Wegener, D. T., MacCallum, R. C., & Strahan, E. J. (1999). Evaluating the use of exploratory factor analysis in psychological research. *Psychological Methods, 4*(3), 272.

Franzen, A., & Vogl, D. (2013). Two decades of measuring environmental attitudes: A comparative analysis of 33 countries. *Global Environmental Change, 23*(5), 1001–1008.

Hedges, L. V., & Olkin, I. (2014). *Statistical methods for meta-analysis*. Academic press.

Hendrickson, A. E., & White, P. O. (1964). Promax: a quick method for rotation to oblique simple structure. *British Journal of Statistical Psychology, 17*, 65–70.

Inglehart, R. F. (2009). Postmaterialist values and the shift from survival to self-expression values. In *The Oxford handbook of political behavior*. https://doi.org/10.1093/oxfordhb/9780199270125.003.0012

Inglehart, R. F. (2018). *Culture shift in advanced industrial society*. Princeton University Press.

Inglehart, R. F. (2020). *Religion's sudden decline: What's causing it, and what comes next?* Oxford University Press.

Inglehart, R. F., & Baker, W. E. (2000). Modernization, cultural change, and the persistence of traditional values. *American Sociological Review, 65*(1), 19–51.

Jorgenson, A. (2012). Are the economy and the environment decoupling? A comparative international study, 1960-2005. *The American Journal of Sociology, 118*(1), 1–44.
Kline, P. (2014). *An easy guide to factor analysis*. Routledge.
Knippenberg, H. (2015). Secularization and transformation of religion in post-war Europe. In *The changing world religion appendix map* (pp. 2101–2127). Springer.
Liddle, B. (2014). Impact of population, age structure, and urbanization on carbon emissions/energy consumption: Evidence from macro-level, cross-country analyses. *Population and Environment, 35*(3), 286–304.
McDonald, R. P. (2014). *Factor analysis and related methods*. Psychology Press.
Morrison, D. F. (1976). *Multivariate statistical methods*. McGraw-Hill.
Mulaik, S. A. (2009). *Foundations of factor analysis*. CRC Press.
Müller, G. P., & Bornschier, V. (1988). *Comparative world data: A statistical handbook for social science*. Johns Hopkins University Press.
Suhr, D. (2012). Exploratory factor analysis with the World Values Survey. In *Proceedings of the SAS Global Forum 2012 Conference, Orlando, FL*.
Tabachnick, B. G., & Fidell, L. S. (2001). *Using multivariate statistics*. Allyn & Bacon.
Tausch, A. (2012). *International macroquantitative data*. Faculty of Economics, Corvinus University of Budapest. Available at http://www.uni-corvinus.hu/index.php?id=47854&tx_ttnews%5Btt_news%5D=0&tx_ttnews%5BbackPid%5D=31638&tx_ttnews%5BcalendarYear%5D=2012&tx_ttnews%5BcalendarMonth%5D=6&cHash=af8ef6888f7c9922b83b113f71c1ca32
Tausch, A. (2019). Migration from the Muslim world to the West: Its most recent trends and effects. *Jewish Political Studies Review, 30*(1–2), 65–225. Available at http://jcpa.org/article/migration-from-the-muslim-world-to-the-west-its-most-recent-trends-and-effects/ (with data definitions and sources). Free data download available from https://www.academia.edu/37568941/Migration_from_the_Muslim_World_to_the_West_Its_Most_Recent_Trends_and_Effects
Tausch, A., & Heshmati, A. (2013). *Globalization, the human condition, and sustainable development in the twenty-first century: cross-national perspectives and European implications*. Anthem Press (with data definitions and sources). Free data download available from https://www.academia.edu/35044095/Globalization_the_human_condition_and_sustainable_development_in_the_21st_Century._Cross-national_perspectives_and_European_implications_Codebook_and_EXCEL_data_file
Tausch, A., Heshmati, A., & Karoui, H. (2014). *The political algebra of global value change: General models and implications for the Muslim world*. Nova Science Publishers.
Yeşilada, B. A., & Noordijk, P. (2010). Changing values in Turkey: Religiosity and tolerance in comparative perspective. *Turkish Studies, 11*(1), 9–27.

Chapter 2
Creative Regulatory Environments: Sandboxes at the Intersection of Regulation and Innovation in an Era of Transition

Sandro Mendonça

Abstract Although there is an ever-increasing demand for innovation (disruption, entrepreneurship, etc.) and regulation (product compliance, economic security), we currently are only starting to understand the conditions for simultaneously stimulating operator and regulator creativity in times of emergent technologies. This work addresses a significant gap in the literature by considering the factors that encourage new knowledge by commercial actors while fostering supervision sophistication. We argue that regulatory sandboxes are place-based institutional devices that allow for pre-adaptive congruence of future technology-rules complexes. We draw from examples from the telecommunications and energy sectors.

2.1 Introduction

Markets are technologically dynamic and regulatory-dense settings. The introduction of products and processes into the marketplace is sometimes thought of, in a stylized way, as giving the defining characteristics to the innovation concept. However, in a modern economy, innovations are not simply thrown into the marketplace; innovators have to navigate a number of norms, rules and laws. Innovators also have to contend and/or comply with the action and determinations of national (e.g. governments, parliaments), sub-national (e.g. provincial and municipal governments) and even supra-national (e.g. International Telecommunications Union, Universal Postal Union, European Central Bank, European Atomic Energy Community, etc.) institutions. This is quite the case in high-tech sectors and network

---

S. Mendonça (✉)
Department of Economics, Iscte Business School, Instituto Universitário de Lisboa (ISCTE-IUL) – Business Research Unit (BRU-IUL), Lisbon, Portugal

UECE/REM – ISEG/University of Lisbon, Lisbon, Portugal

City University of Macau, Macau, China
e-mail: sfm@iscte-iul.pt

© The Author(s), under exclusive license to Springer Nature Switzerland AG 2024

T. Devezas et al. (eds.), *Global Energy Transition and Sustainable Development Challenges, Vol. 1*, World-Systems Evolution and Global Futures,
https://doi.org/10.1007/978-3-031-67583-6_2

industries. When safety risks and externality effects are pervasive, active on-going regulation is part and parcel of techno-economic development.

Governance mechanisms are, therefore, very much part of the innovation process. New products and processes hit the market, but also a thick cobweb of regulatory dispositions and deliberations. For a long time, regulation was thought as a barrier and a source of perverse distortions; the final goal of good regulation was, therefore, to contribute to its own extinction. Regulation, nevertheless, provide services to incumbents and insurgents, to established industries and newcomer niches. Indeed, in high-tech network fields, regulatory agencies often act in ways that go beyond simply correcting market failures.

This chapter departs from the observation that regulators can themselves, be considered part and parcel of innovation systems. Though often little emphasized, regulators (individually or in association with others) do generate innovations, use innovations by the private sector, and shape changes in technologies and businesses. This implies that it would be hard to conceive regulators to be passive in times of digital and energy transition, that is, when breakthrough technologies and novel business models emerge in the context of several societal challenges.

In this contribution, we show how 'regulatory sandboxes' are policy devices that work at the intersection of regulation and innovation processes. We call them an instance of 'Creative Regulatory Environments'. Sandboxes are a recent instrument in national administrative authorities' toolbox entrusted with regulating markets, supervising sectors and surveilling products or infrastructures. We argue that these secure spaces represent a new service by sectoral regulators to the realm of the market economy. Many benefits can be perceived as being brought about by this solution, including lowering uncertainty for operators and fostering learning by regulators. The outcome may well be a more aligned but also more vibrant (re)mix (ed) economy.

In what follows we start by surveying the original rationale and historical drifts of regulatory activity. We go on to present what are sandboxes, what they mean for strategy and how they can be rationalized as 'creative regulatory environments'. We then provide illustrations from the communications and the energy areas. We end by highlighting the importance of experimentation approaches to innovation policy and regulatory strategy.

2.2 Shifting Regulatory Landscape

2.2.1 Wave of Deregulation

Whereas before unimpeded competition was deemed to be potentially destructive in terms of quality and reliability, by the 1970s a new consensus was forming that regulation was suppressing innovation, sheltering inefficiency, denying variety and leading to severe misallocation of resources (Kahn, 1988, p. xvi). Indeed, ineffective regulation, incomprehensible rules, corruption-prone bureaucracies, and the cost of compliance have since then become associated with the notion of regulation.

By the early 1980s the Keynesian regime of state intervention at the macro-economic was being dismantled at the domestic level while capital flows and trade competition were rising at the international level. The process of marketization known as the 'Washington Consensus' (a term coined in the late 1980s) was characterized by privatization, liberalization and deregulation but also by a renewed de-emphasis on planning and industrial policy, a replacement of fiscal policy with incentives design, and a reassertion of property rights, especially after the fall of the Berlin Wall (see Williamson, 2009).

2.2.2 Rise of Regulators

The quest for deregulation was felt at the sectoral level in areas such as telecommunications, energy and civil aviation. However, this process was accompanied by establishing a new set of institutions to protect and promote competition (Hermann and Flecker, 2012). Privatization of public utilities, liberalization of economics and the deregulation of markets came to be managed by independent specialized regulatory agencies, which by the 1990s had developed into a recognizable of economic organization and a key vehicle for a new mode of governance in the economic system (Kahn, 2003; Majone, 1997). These actors were entrusted with 'quasi-legislative' and 'quasi-judicial' powers so as to keep a high level of competitive pressure when market forces are too feeble to guarantee it (Cuéllar, 2001).

The trend towards lowering administrative constraints coincided with a growth in the efforts in standard harmonization and mutual recognition of rules across countries, often with a drive for higher environmental protection, public health, consumer safety and technical interoperability at the supra-national and global levels (Majone, 1993; Chayes and Chayes, 1995). In other words, 'regulatory reform' ended up being a combination of *deregulation* and *re-regulatory* trends (Ménard and Ghertman, 2009). The 'post-regulatory' world turned out to be a complex one.

2.2.3 Regulation Reviewed

By the 2000s the agenda was going beyond 'regulate and forget', it was indeed turning to issues of 'regulatory management' and 'regulatory quality' (see OECD, 2002). Under the guise of labels like 'better regulation' or 'smart regulation' greater emphasis was laid on review exercises and the use of 'new public management' methods (see, e.g. Govaere and Garben, 2018). Tools have ranged from the introduction of regulatory impact assessment to the digitalization of processes (Malyshev, 2008). By the 2010s, after years of public utility privatization, some agencies had become quite sizeable and no longer dealt with the historical incumbents (Decker, 2015, p. 7).

2.2.4 *Anticipatory, Adaptive, Agile Regulation*

The most recent experiences of modern regulation have moved beyond supervising the restructuring of industries from monopoly to competition to include more dynamic concerns. By the turn of the 2020s the regulatory debate evolved from a to an 'assess then regulate', to an 'anticipate but adapt' mode of regulatory posture. The author witnessed these shifts during his professional experience in regulation, which included participating at the Network of Economic Regulators at the OCDE, by then a prime locus for benchmarking and mutual learning among experts and decision-makers (see also OECD, 2021).

An emergent posture of regulators can be inferred from a number of tools and attitudes: facilitating new forms of inter-firm arrangements (e.g. infrastructure sharing, energy communities), encouraging technology transitions (from 4G to 5G, low-carbon power, etc.), managing sectoral decline (universal postal service), ensuring stability (in the face of pandemics, of inflation bursts, etc.), using non-coercive signals (behavioural nudging), encouraging increased consumption (of connectivity and the benefits associated with data), enhancing stakeholder engagement (consumers, women, civil society associations), etc.

2.3 Regulation in Dynamic Settings

2.3.1 *From Static Rivalry to Dynamic Competition*

Government regulation was originally justified on the basis of negative externalities, like pollution, that had to be curbed either through commands or incentives. Direct regulation by dedicated agencies refers to cases in which firms are made to comply with specific rules or guidelines, for instance, access to radiospectrum rights by telecom companies or permission to inject electricity from a dam by energy companies.

The most relevant form of competition is not just among existing firms trying to beat the others by yielding price or capacity advantages. Usually, competition in contemporary economies involves some degree of 'dynamic competition', that is, 'potential competition from new products and production processes' that may displace current product or production processes (Cabral, 2017, p. 382).

2.3.2 *Regulating in a Fast-Moving Economy*

When innovation is taken into consideration, the regulatory challenge becomes more complex. For instance, issues of 'merit' have percolated into anti-trust/competition policy. Decisions about anticompetitive conduct and merger control are no longer exclusively made on static concepts and short-term criteria such as price competition and market concentration. The mainstream view coalesced around the notion that disruptive firms, either big or small, that challenge the status quo are to be welcomed in this area and be allowed to exploit their leadership (Federico et al., 2020). The question is important for sectoral regulation since, with dynamic markets, it is not satisfactory to rely alone on standard anti-trust principles (de Streel, 2008, p. 727).

Innovation in sectoral regulation should come as more natural than in competition policy. For instance, several innovations and new technical capabilities have fundamentally changed telecommunications since the 1980s. Part of the transformation was a set of reforms that led to a virtuous cycle of increased competition, increased investment, price declines, and accelerated adoption of information and communications technologies (ICTs) (Cave et al., 2019). Another example is the structural reorganization of energy markets, whereby the arrangements around the transmission grid shifted momentously in several markets (Glachant et al., 2008). The relevant literature is still scant.

2.3.3 *When Regulators Innovate*

Regulation has been recognized as a non-market factor in leading innovation and influencing diffusion (Beise and Rennings, 2005). Regulations create pressures and opportunities that can induce innovation. Different regulations and different implementation types of a given regulation can generate innovation impacts in various ways (Blind, 2012). The most recent question is how regulation officials can 'best encourage welfare-enhancing innovation' (Braunerhjelm et al., 2023). Our topic in this contribution is how sandboxes fit this picture. In this context we will understand the roll-out of sandboxes as a 'regulatory innovation', that is, a new solution for regulatory functions, organizational processes and institutional structures (Black et al., 2006).

Regulation cannot focus solely on consumer rights vis-à-vis today's offerings or on the cost-efficient use of existing infrastructure. Regulators are increasingly witnessing fast structural changes and sometimes even face the very definition of the sectors they supervise melt away. One way to adjust has been by giving more attention to not creating unjustified barriers against innovation (for instance, the Portuguese authority for communications—ANACOM—and the Brazilian agency for energy—Aneel—have created departments for dealing with innovation) and to consider the temporary removal of regulatory burdens to let regulatory lessons to emerge in real time (on the way to allow for this have been sandboxes).

2.4 Playing in Sandboxes, Playing with Sandboxes

2.4.1 Introducing Regulatory Sandboxes

The balance between welcoming innovation and protecting users is difficult but critical in today's world. It is usually acknowledged that the UK's Financial Conduct Authority first introduced a 'regulatory sandbox' for nurturing 'fintechs' in November 2015 (Cornelli et al., 2020). The idea, which was around by the early 2010s, was to 'promote competition by supporting disruptive innovation' (Wechsler et al., 2018).

In April of the following year, the *Financial Times* 'FCA does a big number to prove it is the font of financial wisdom', April 5, 2016) reported:

> The balance between allowing innovation to thrive and protecting customers is a tough one to achieve but it is a critical one. Too much regulation and the industry becomes sclerotic, driving entrepreneurs to Silicon Valley and bogging down consumers with antiquated systems and products. Too little, and fraud abounds. As a concept, financial innovation does not have the best of reputations.
>
> To this end, the FCA's proposed "regulatory sandbox" is intriguing. The idea is that it allows companies to experiment with new products in a controlled environment.

The concept would soon be followed in other countries, and afterwards in other industries (Wechsler et al., 2018; Cornelli et al., 2020) (see Fig. 2.1). As an addition to the regulatory toolbox, sandboxes have become quite recognized as a way to speed up regulation in the face of emerging technologies (NESTA, 2019) but also increasingly popular beyond the most advanced economies (IADB, 2020).

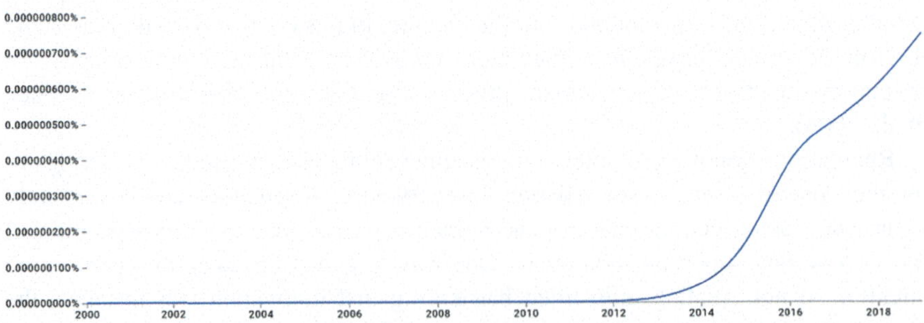

Fig. 2.1 The frequency of 'regulatory sandbox' in printed materials. Source: Google ngrams

2.4.2 Sandboxes as a Vehicle for Experimental Strategizing

Experimental approaches have increasingly been deployed by regulators and other government entities to better understand the actual workings and impacts of rules and remedies. The need to take experimentation seriously signals the acknowledgement of regulators' bounded rationality and of the genuinely surprising features of structural change. Importantly, the strategic rise of experimentation in regulatory priorities is related to the need to gain actual experience that could be used to inspire new regulation or revise existing supervision techniques.

Behavioural experiments have been one approach and such tentative interventions have been used by Ofgem (the UK's energy regulator), the Danish Competition and Consumer Authority and the Dutch Authority for Consumers & Markets (CMA, 2023). *Regulatory experimentation* has been another approach, and they have often taken the form of derogations, waivers or exemptions from a rule while letting players to conduct activities that they would not normally be allowed to engage in without explicit a priori commitments to future authorizations (Vaseli et al., 2021).

Sandbox set-ups emerged as a way to standardize, give transparency and enshrine as strategy ad hoc or 'muddling through' practices of conducting regulatory policy. These costume-made spaces provide a flexible room to innovation proponents and give room for prospective markets but under the close watch of supervisors. They allow both for innovation facilitation and regulatory responsiveness. Thereby, a reason behind the enthusiasm for sandboxes is that they have the potential to reduce information asymmetries for regulators but also uncertainties for the regulated while attenuating unfair gains by some operators due to the regulatory lag phenomenon.

2.5 Creative Regulatory Environments

2.5.1 How to Understand CREs

In this contribution, we capitalize on innovation studies literature to understand regulatory sandboxes as 'creative regulatory environments' (or CREs). We start by observing that an undertaking's (or group of players) creative and entrepreneurial potential depends substantively on the surroundings in which that individual, corporate or industry develops its activity (see, e.g. Hemlin et al., 2008). We advance the concept of CRE to study new phenomena of regulatory experimentation and, in particular, to gain perspective on how a sandbox influences technology and business innovation. We set forth a definition of CRE:

> CREs are those places and contexts, defined by a precise perimeter and explicit guidelines, that enable individuals and organizations to experiment beyond what would be permissible in the market, thereby exerting an accelerating and pre-adaptive influence on undertakings engaged in innovation and technology-based work aiming to produce or test new operational knowledge or business models.

First, no innovation is expected to continue operating in a normative void after its launch. A sandbox intentionally creates an artificially safe space for innovation. Hence, they can be considered a delimited 'legal enclave' (Philipsen et al., 2021) in the broader 'macro-environment' (Caraça et al., 2009) that regulatory institutions create around the innovation process. Once established, this institutional device creates new difficulties for established regulators in terms of certainty, equality and accountability that will need to be resolved and accommodated (see Radosevic et al., 2023).

Second, not only are CRE bounded in (physical and/or virtual) space, but they are also managed with attention to time. Dedicated places for conducting trial and error exercises are developmental contexts akin to 'temporary knowledge ecologies' (Bathelt and Zeng, 2015). That is to say, in these sites there is a mobility of participants and a concentrated relational dynamics in a way that is similar to hackathons or trade fairs. CREs work as unique events or succession of project-based activities that are rich in knowledge-based inputs and help configure entire fields.

Third, in moving from static to dynamic regulation, CRE signify the incorporation of learning as a constitutive process of coming up with new visions of markets. Participants are atypical (forerunners, innovators) and test samples (unfinished the product) of their capabilities. If experiments are successful, the regulator can use the lessons to inform future frameworks. In the case of sandboxes, both 'regulatory pilots' (innovative practices are tested) and 'pilot regulation' (new rules are tested) can be experimented with while keeping entrepreneurial risk and regulatory effort within bounds (Schittekatte et al., 2021).

2.5.2 Regulators Supporting Discovery through CREs

CREs give 'serious play', that are prototyping and demonstrations, a central standing in the regulatory process (see Dodgson et al., 2005). It can be integrated with standard regulatory practices, which in the case of telecommunications already admit special temporary licences for experimental uses of spectrum resources to improve innovative wireless technologies and their management (see Bustamante et al., 2017).

By pushing forward CREs, governments and regulatory agencies foster a new way to conduct innovation-diffusion policy. In particular, regulatory sandboxes play a role in the State-Market interaction. This tool gives 'voice' to both sides by empowering both sectoral authorities and interested parties (operators, vendors, consultants, universities, etc.). It gives public and private actors a context to increase legitimacy, to better risk management capabilities, to share knowledge, to unchain serendipity, to avoid premature lock-ins, and to develop foresight in an open but controlled way (see Alaassar et al., 2020; Audretsch and Keilbach, 2008; Cunha et al., 2010; Santos et al., 2021; Leitão and Ratten, 2022).

2.5.3 *CRE-Empowered Regulation and Innovation Systems*

When instituting and running CREs, regulators and other public entities are mindful that the short-term impacts of regulations are often negative for innovation (and can be minimized by mitigating risks), in contrast to their often positive long-term implications (which can be maximized by enhancing future demand) (Blind, 2016). Regulators in such environments are more amenable to provide advice than to exercise their punishment duties, thus pre-empting future attrition and litigation. CRE can thus be seen to strengthen the scope for participation in regulation (see Mendonça, 2021).

The modulating and mediating role of regulation suggests a holistic outlook in innovation governance, that is to say, regulators do have a place as recognizable institutions in national (and international) systems of innovation (Fagerberg, 2016). Through CREs regulators participate in the innovation process, however not in the prescriptive formal Science-Technology-Innovation (STI) mode but in the more pedestrian Doing-Using-Interacting (DUI) mode, which relies on informal processes of problem-solving experience-based know-how development (Jensen et al., 2007). A DUI mode for regulators promotes learning within and across organizations and can be argued to have a stronger impact on non-technological innovation, namely in being an external source of learning for firms in managing new-to-market products and procedures in a more responsible way before they hit their final costumers (Parrilli et al., 2016; Haus-Reve et al., 2023).

2.5.4 *Dialectics of Trade-Offs and Trade-Ins Yielding a Bias for Dynamism*

CRE practices, like regulatory sandboxes, encourage regulatory decision-makers to balance boldness and caution. Regulators are thus encouraged to remain evidence-based and future-oriented in their diagnoses of what works and what does not work, while not embarking on 'newism' (the notion that innovation is exclusively virtuous) and by not all too quickly condemning situations perceived as harmful (such as ambiguity) (see Quan, 2020, and Lambert 2022, p. 66). Shared cognition and contextually embeddedness in sandboxes can be pragmatically mobilized to move forward amid puzzles and wicked problems.

Sandboxes, as a special case of CREs, can thus be seen as 'regulation-by-paradox'. Innovation and regulation may be contradictory at first look, yet they are interrelated elements that exist simultaneously and co-evolve over time in modern economies (see Cunha and Putnam, 2019). CREs allow for handling disequilibrium to move from 'either-or' to 'both-and' thinking. Tensions can be recognized in the need to reconcile contingency with routine (i.e. situational decisions vs. regulatory predictability) and knowledge development with prerequisites of independent regulation such as transparency, proportionality and non-discrimination.

With CREs regulators can embrace surprises and take advantage of the trade-offs (e.g. product features that may benefit one stakeholder at the expense of others, like firms against consumers) and the trade-ins (synergies that may unexpectedly provide a way to surmount) that result from navigating paradoxes. CREs help the ongoing process of *regulating* and yield a bias for dynamism in *regulation*.

2.6 Two Experiences with Sandboxes in Telecom and Energy

2.6.1 *The Portuguese Navy and the Portuguese Model of Sandboxes*

In Portugal the Infante D. Henrique (IDH) was the Free Technology Zone (ZLT) to be established in 2022. It is headquartered at CEOM (Maritime Operational Experimentation Centre), a built environment that the Portuguese Navy decided to reuse between a bay and the ocean off Setúbal (Fig. 2.2). Here, the Navy aims to nurture collaboration with industry, academia and international cooperation with the aim of testing emerging systems and technologies, use of unmanned vehicles, new materials and artificial intelligence in a real environment and in the open sea.

The Navy's ZLT was the first to be established after the Decree No. 67/2021, published in July 30 of 2021, which instated the legal regime defining the governance model for the promotion of innovation through the creation of ZLTs. The ZLT notion draws from the 'regulatory sandbox' concept. The idea was first floated during the year 2018 by the Portuguese government and loosely articulated as 'is

Fig. 2.2 The original site of the IDH. Credits: Marinha Portuguesa; composition by the author

Fig. 2.3 The role of the communications regulator. Credits: The author

the physical or regulatory space that enables experimentation of new technologies and the attraction of investments for testing initiatives' (*Diário de Notícias*, 'Boas ideias para o futuro', April 9, 2018). Actual work started behind the ZLT model started after the government's formal pronouncement RCM 29/2020, which referred to the need to develop a flexible and harmonized approach to emerging technologies like electric mobility, Internet-of-things, etc.

In the Portuguese model, the National Agency for Innovation (ANI) acts as coordinator and manager of the ZLT network, and liaises with the national regulatory bodies. When the Navy's ZLT was inaugurated, there were other ZLTs under analysis. In the end, the Decree of 2021 stipulates that ZLTs are:

> ... test sites, geographically located, intended for testing of innovative technology, products, services and processes, by their promoters, in a safe manner, with the support and monitoring of the respective competent authorities, namely in terms of testing, provision of information, guidelines and recommendations, corresponding to the concept of regulatory sandbox.

The bylaws of the IDH were drafted as a collaborative effort with multiple actors, specifically the national testing authority (a role carried out by ANI), but also regulators, namely ANACOM (communications), AAN (aeronautics) and ANAC (civil aviation). Other entities were involved in the process, in particular the Defense Ministry and the Ministry of Transition and Climate Action.

It is at IDH that the Navy has organized the *Robotic Experimentation and Prototyping augmented by Maritime Unmanned Systems* (REPMUS) exercise. A REPMUS a wide number of national and international entities participating from industry, academia, and the foreign navies—it immediately became the largest NATO's experimentation project. In 2023 (2 weeks in September), 1400 participants were engaged, 26 navies were present, and over 80 aerial, surface and subsurface drones were used.

In the REPMUS exercise the Portuguese communications regulator has been heavily involved throughout (since the exercise of 2021; see Fig. 2.3). Every year, before the event, ANACOM is included in the planning by the Navy. As the event approaches, it allocates radiofrequencies to the participants (indeed, it was prompted to develop in-house a new software application to deal with the increased pressure of radiospectrum demand). During the event, ANACOM monitors the airwaves and contributes to debugging if interferences or incidents occur. After the event

ANACOM analysis the readings it took from the event, including the patterns of electromagnetic emissions.

2.6.2 Sandbox Initiatives by the Brazilian Energy Regulator

In December 2021, Aneel published the reference terms for pilots of electricity tariffs. This First Call for Tariff Sandboxes attracted 14 within-scope proposals, submitted by electricity distributors or their business groups. After evaluation, six proposals for tariff sandboxes were approved in April 2023 (Fig. 2.4).

Annel's idea was to invite pilot-projects in the form of temporary authorizations for tariff schemes (or new ways of billing energy consumption) within the bounds and criteria defined by the agency. Sandboxes could test solutions such as pricing schemes, smart meters, and pre-pay options. Sandbox proposals had to involve customers from the outset. Then, sandboxes could not impact the current price policy of operators.

It should be noted that the regulator made a reference to the 'Open Innovation' approach since promoters could build a network for sharing experiences (especially with consumers). Financial support was budgeted through the R&D funds available at the regulator. By July 2023 Aneel expanded its vision and made an open call for the energy sandbox design, of which tariff sandboxes will be just one example.

Fig. 2.4 The first initiative for tariff sandboxes at Aneel. Credits: Aneel's website

2.7 Conclusions

Public regulation, not only private innovation, is a learning process. Sectoral authorities and supervision agencies need to countervail asymmetric information (and asymmetric knowledge). Sandboxes are examples of creative and catalysing places where beta-products and beta-regulations can co-emerge and co-evolve, thereby fostering learning before the actual offerings and technologies are formally introduced.

Regulatory experiments in sandboxes are a leading example of 'creative regulatory environments' (CREs). In such settings, while embracing a multidisciplinary perspective, what goes on is regulation-by-paradox, that is to say, regulators act in a context of disequilibrium and intervene through a dialectic, iterative and interactive but also didactic, advisory and dialogical style of regulation. CREs are prudential and entrepreneurial spaces construed to achieve dynamic but balanced outcomes that can contribute to the common good.

Recent cases from communications in Portugal and from energy in Brazil illustrate the main points of this text. The Portuguese case has a top-down history and is circumscribed to one place, while the Brazil case is bottom-up and distributed across several regions. ANACOM played a central role in the first sandbox in Portugal, which was driven by the Navy and is focused on tech-based innovation. Aneel led a project to unleash tariff schemes proposals from firms that had to include customers in the exercises. The cases also show how regulatory sandboxes may be customized in a variety of ways.

Acknowledgements The work leading to this version was presented at the Join Research Centre, Seville. I thank Dimitrios Pontikakis and all the participants. In the Portuguese Navy I Thank Admiral Vicente Rodrigues and Commanders Lourenço Piedade, António Mourinha and Paulo Simões. At Annel I thank Paulo Luciano and Alex Cavalcante Alves. Comments by Ana Teresa Santos, Joana Nunes and Telmo Santos are greatefully acknowledge. Finally, I thank João Leitão for the opportunity to collect my impressions in the form of this chapter. The author refers to the support of grants UID/GES/00315/2013, UIDB/00315/2020; UIDB/05069/2020; PTDC/EGE-ECO/30690/2017 and is part of the project PTDC/EGE-ECO/30690/2017.

References

Audretsch, D. B., & Keilbach, M. (2008). Resolving the knowledge paradox: Knowledge-spillover entrepreneurship and economic growth. *Research Policy, 37*(10), 1697–1705.

Alaassar, A., Mention, A.-L., & Aas, T. H. (2020). Exploring how social interactions influence regulators and innovators: The case of regulatory sandboxes. *Technological Forecasting and Social Change, 160*, 120257.

Bathelt, H., & Zeng, G. (Eds.). (2015). *Temporary knowledge ecologies: The rise of trade fairs in the Asia-Pacific region*. Elgar.

Black, J., Thatcher, M., & Lodge, M. (Eds.). (2006). *Regulatory innovation: A comparative analysis*. Elgar.

Beise, M., & Rennings, K. (2005). Lead markets and regulation: A framework for analyzing the international diffusion of environmental innovations. *Ecological Economics, 52*(1), 5–17.

Blind, K. (2012). The influence of regulations on innovation: A quantitative assessment for OECD countries. *Research Policy, 41*(2), 391–400.

Blind, K. (2016). The impact of regulation on innovation. In J. Edler, P. Cunningham, A. Gök, & P. Shapira (Eds.), *Handbook of innovation policy impact* (pp. 450–482). Elgar.

Braunerhjelm, P., Andersson, M., Blind, K., & Eklund, J. (Eds.). (2023). *Handbook of innovation and regulation*. Elgar.

Bustamante, P., Weiss, M., & Sicker, D. (2017). Exploration of the Federal Communications Commission's experimental radio service (ERS): Understanding ten years of experimental spectrum licenses. *SSRN 2943486*.

Cabral, L. (2017), Introduction to Industrial Organization, second edition, Boston, Mass.: The MIT Press.

Caraça, J., Lundvall, B.-Å., & Mendonça, S. (2009). The changing role of science in the innovation process: From Queen to Cinderella? *Technological Forecasting and Social Change, 76*(6), 861–867.

Cave, M., Genakos, C., & Valletti, T. (2019). The European framework for regulating telecommunications: A 25-year appraisal. *Review of Industrial Organization, 55*, 47–62.

Chayes, A., & Chayes, A. H. (1995). *The new sovereignty: Compliance with international regulatory agreements*. Harvard University Press.

CMA. (2023). *Experiments at the CMA: How and when the CMA uses field and online experiments*. CMA.

Cornelli, G., Doerr, S., Gambacorta, L., & Merrouche, O. (2020). *Regulatory sandboxes and fintech funding: Evidence from the UK*. BIS Working Paper No. 901.

Cuéllar, L. (2001). *As Agências Reguladoras e Seu Poder Normativo*. Dialética.

Cunha, M. P., Clegg, S. R., & Mendonça, S. (2010). On serendipity and organizing. *European Management Journal, 28*(5), 319–330.

Cunha, M. P., & Putnam, L. L. (2019). Paradox theory and the paradox of success. *Strategic Organization, 17*(1), 95–106.

Decker, C. (2015). *Modern economic regulation: An introduction to theory and practice*. Cambridge University Press.

Dodgson, M., Gann, D., & Salter, A. (2005). *Think, play, do: Technology, innovation, and organization*. Oxford University Press.

Fagerberg, J. (2016). Innovation policy: Rationales, lessons and challenges. *Journal of Economic Surveys, 31*(2), 369–659.

Federico, G., Morton, F. S., & Shapiro, C. (2020). Anti-trust and innovation: Welcoming and protecting disruption. *Innovation Policy and the Economy, 20*(1), 125–190.

Glachant, J.-M., Dubois, U., & Perez, Y. (2008). Deregulating with no regulator: Is the German electricity transmission regime institutionally correct? *Energy Policy, 36*(5), 1600–1610.

Govaere, I., & Garben, S. (Eds.). (2018). *The EU better regulation agenda: A critical assessment*. Bloomsbury Publishing.

Haus-Reve, S., Fitjar, R. D., & Rodríguez-Pose, A. (2023). DUI it yourself: Innovation and activities to promote learning by doing, using, and interacting within the firm. *Industry and Innovation, 30*(8), 1008–1028.

Hermann, C., & Flecker, J. (Eds.). (2012). *Privatization of public services: Impacts for employment, working conditions, and service quality in Europe*. Routledge.

Hemlin, S., Allwood, C. M., & Martin, B. R. (2008). Creative knowledge environments. *Creativity Research Journal, 20*(2), 196–210.

IADB. (2020). *Regulatory sandboxes and innovation testbeds: A look at international experience in Latin America and the Caribbean*. Washington, D.C.

Jensen, M., Johnson, B., Lorenz, E., & Lundvall, B. A. (2007). Forms of knowledge and modes of innovation. *Research Policy, 36*, 680–693.

Kahn, A. E. (1988). *The economics of regulation principles and institutions* (Vol. I). The MIT Press.
Kahn, A. E. (2003). *Lessons from deregulation telecommunications and airlines after the crunch*. Brookings Institution.
Lambert, G. (2022). James March: A postmodern perspective on organization without management theory. *Journal of Management History, 28*(1), 66–88.
Leitão, J., & Ratten, V. (Eds.). (2022). *Context-based entrepreneurship: The importance of location, time, and culture*. Springer.
Majone, G. (1993). *Deregulation or re-regulation? Policymaking in the European Community since the Single Act*. European University Institute working paper No.93/2.
Majone, G. (1997). From the positive to the regulatory state: Causes and consequences of changes in the mode of governance. *Journal of Public Policy, 17*(2), 139–167.
Malyshev, N. (2008). *The evolution of regulatory policy in OECD countries*. OECD.
Ménard, C., & Ghertman, M. (Eds.). (2009). *Regulation, deregulation, reregulation institutional perspectives*. Elgar.
Mendonça, S. (2021). Dinâmica e estrutura das consultas públicas: A regulação das comunicações em Portugal, 1998–2020. *Revista da CGU, 13*(24), 269–283. IADB.
NESTA. (2019). *Renewing regulation: 'Anticipatory regulation' in an age of disruption*. NESTA.
OECD. (2002). *Regulatory policies in OECD countries: From interventionism to regulatory governance*. OCDE.
OECD. (2021). *OECD regulatory policy outlook 2021*. OECD.
Parrilli, M. D., & Heras, H. A. (2016). STI and DUI innovation modes: Scientific-technological and context-specific nuances. *Research Policy, 45*(4), 747–756.
Philipsen, S., Stamhuis, E. F., & de Jong, M. (2021). Legal enclaves as a test environment for innovative products: Toward legally resilient experimentation policies. *Regulation & Governance, 15*, 1128–1143.
Quan, D. (2020). *A few thoughts on regulatory sandboxes*', Technical report,. Stanford PACS, Stanford University.
Radosevic, S., Kanellou, D., & Tsekouras, G. (2023). The experimentation–accountability trade-off in innovation and industrial policy: Are learning networks the solution? *Science and Public Policy, 50*(4), 655–669.
Santos, A. B., Bogers, M. L., Norn, M. T., & Mendonça, S. (2021). Public policy for open innovation: Opening up to a new domain for research and practice. *Technological Forecasting and Social Change, 169*, 120821.
Schittekatte, T., Meeus, L., Jamasb, T., & Llorca, M. (2021). Regulatory experimentation in energy: Three pioneer countries and lessons for the green transition. *Energy Policy, 156*, 112382.
de Streel, A. (2008). Current and future European regulation of electronic communications: A critical assessment. *Telecommunications Policy, 32*(11), 722–734.
Veseli, A., Moser, S., Kubeczko, K., Madner, V., Wang, A., & Wolfsgruber, K. (2021). Practical necessity and legal options for introducing energy regulatory sandboxes in Austria. *Utilities Policy, 73*, 101296.
Wechsler, M., Perlman, L., & Gurung, N. (2018). The state of regulatory sandboxes in developing countries. *Available at SSRN*.
Williamson, J. (2009). A short history of the Washington Consensus. *Law and Business Review of the Americas, 15*(1), 7–23.

Chapter 3
Examining the Behaviour of the Kuznets Curve Under Economic Policy Uncertainty Conditions Using the Stochastic Frontier Analysis: Evidence from 11 Selected Countries

V. Moutinho, J. Leitão, and B. Henriques

Abstract This work aims to investigate and validate the Kuznets relationship between carbon dioxide emissions and economic drivers. More specifically, it considers Gross Domestic Product per capita, Transparency Corruption, Revenues from Natural Resources and Renewable Consumption under Economic Policy Uncertainty conditions for 11 countries in two groups, namely the G7 and BRIC, from 1997 to 2020. A stochastic frontier analysis was conducted to validate a functional translog for the environmental Kuznets curve (EKC), considering time-varying and time-invariant models. The results of the stochastic frontier show that for the EKC, the Inverted U Shape form is valid for the relationship between total resource rents and CO_2 emissions per capita, and the relationship between EPU (uncertain economic policy) and CO_2 emissions per capita, according to the results of the one frontier varying model. Regarding the influence of renewable energy consumption on CO_2 emissions per capita, the Inverted U Shape form is validated according to the results of both of the two time-varying boundary models. Alternating with the U-shaped form, the relationship between the Corruption Index and the per capita CO_2 emissions is validated according to both of the two time-varying models.

V. Moutinho (✉) · J. Leitão
NECE – Research Center for Business Sciences and Management and Economics Department, University of Beira Interior, Covilhã, Portugal
e-mail: ferreira.moutinho@ubi.pt; jleitao@ubi.pt

B. Henriques
Management and Economics Department, University of Beira Interior, Covilhã, Portugal
e-mail: b.nunes.henriques@ubi.pt

© The Author(s), under exclusive license to Springer Nature Switzerland AG 2024

T. Devezas et al. (eds.), *Global Energy Transition and Sustainable Development Challenges, Vol. 1*, World-Systems Evolution and Global Futures,
https://doi.org/10.1007/978-3-031-67583-6_3

3.1 Introduction

The environmental Kuznets curve (EKC) illustrates the connection between economic growth and the extent of pollution or environmental deterioration. Although this relationship can sometimes be presented in controversial terms in the literature, it has been the subject of increased attention, especially in terms of analysing the causal relationship between levels of economic growth and CO_2 emissions for groups of countries with different levels of development. However, this relationship is not straightforward and there is still an open debate in the literature between those who argue that economic growth harms the CO_2 emissions of the countries under study and those who argue that there is a positive relationship.

Over the last decade, economic systems have shown behaviour that oscillates between periods of high and low volatility. Currently, the relationship between the behaviour of economic growth and the conditions of countries' climate change policies cannot be ignored. Furthermore, technological advances and structural changes, as well as the importance of energy and the environment, have become highly important issues at a global level, always based on the eco-efficiency of countries over the years. Therefore, access to and acquisition of raw materials and materials for the production of products that incorporate technology may have been compromised over time. As a consequence, rapid changes in international markets with serious impacts on energy prices are associated with uncertain changes in economic policy. They have adverse effects on domestic economies whose dependence on energy, natural resources and raw materials, particularly in international markets, can condition and jeopardize the goals and commitments made to mitigate polluting gases. We must therefore consider that in global economies, particularly for several industrialized countries, economic uncertainty policies affect public and private investment decisions. Over time, there have been numerous environmental concerns and their link to countries' economic growth. Understanding how the two variables are linked goes beyond the well-known environmental Kuznets curve, requiring the incorporation and empirical testing of other variables: economic (namely the conditions of economic policy uncertainty); environmental (the consumption of renewable and non-renewable energy); and institutional (corruption and public debt), in addition to economic growth and CO_2 emissions.

Regarding the conditions of economic policy uncertainty, according to the author (Zeng and Yue 2022), this index is considered significantly important for the society of countries since this index is controlled by policymakers and can sometimes affect the economy from a macro- and microeconomic point of view. This author also argues that in countries with high values on the Economic Policy Uncertainty Index, investors tend to be more "conservative", consumption tends to be lower as well as investment and possibly a lower level of debt issuance. The Economic Policy Uncertainty Index developed by Baker et al. (2016) advocates that high levels of this index in countries can harm their macroeconomic performance and affect the volatility of share prices, investment values and employment growth.

As its name suggests, Renewable Energy Consumption corresponds to the value of energy consumption allocated to renewable energy sources, such as rain, wind, sea, sun and even geothermal energy. Over time, according to Sachan et al. (2023), the environment has always been society's top priority, and awareness of the need to use more renewable energy sources has grown internationally. Even so, there is still a high level of consumption of non-renewable variables around the world and, as such, it is pertinent to include this variable in the work. Since rapid economic growth has been the cause of excessive energy consumption and, consequently, a huge increase in greenhouse gas and CO_2 emissions, the environmental issue has become more noticeable (Ali et al., 2023).

Concerning institutional variables, corruption, according to García (2019), does not have a comprehensive and universally accepted definition, yet this author advocates that corruption is an act of abuse of power to achieve a certain private gain. When it comes to the relationship between corruption and a country's economic growth, there are authors who argue that it is harmful and corrosive to a country's economy, such as Uberti (2021). However, when analysing its relationship with environmental degradation, the conclusions are bidirectional. Since there are positive relationships, i.e. more corruption leads to greater environmental degradation, as Mehmood (2022) studies, it is also possible to see significantly negative relationships.

As far as corruption is concerned, according to Transparency International, the most corrupt countries tend to be less developed countries, particularly poorer countries located mostly in Africa or Asia. On the other hand, more developed countries are less corrupt, such as Denmark, Sweden and even Germany. According to the same source, countries that fail to combat high levels of corruption worsen the effects of the climate crisis. As far as the Economic Policy Uncertainty Index is concerned, according to the EPU's website, this index tends to increase at the time of landmark events such as the Second Gulf War, the Euro Debt Crisis, etc. For this reason, the highest values tend to occur in more advanced economies with closer trade and financial ties or as a result of the high level of instability experienced in each country. When it comes to energy consumption, it is not possible to approach the subject by dividing the conclusions between developed and developing countries, because renewable energy consumption depends on the resources coming from each country. According to Wisevoter, countries like Sweden, Brazil, Denmark and even Colombia have a highly renewable energy matrix. On the other hand, countries such as Russia still have a low percentage of Renewable Energy Consumption in their daily lives. From a different perspective, the consumption of fossil fuels remains substantial in countries such as China, the United States of America, India, Russia, and even Japan, as reported by the official Statista website.

Therefore, we must consider that in global economies, particularly for a set of industrialized countries, economic uncertainty policies affect public and private investment decisions. On the other hand, economies need to meet the challenges and commitments within the Kyoto Protocol, including the decision to expand renewables and reduce the production of non-renewable energy, so that the energy matrix and its composition affect the processes and the entire product value chain.

However, the more industrialized economies are still faced with an uncertainty of economic policy that directly affects the energy carrier and, consequently, carbon emissions. The literature often overlooks the role of political uncertainty, particularly in its significance when analysing the correlation between economic growth and greenhouse gas emissions.

3.1.1 Contextual Setting

In terms of the sample, this study considers two different groups, totalling 11 countries, namely the G7 group and the BRICS, made up of the world's major economic powers. It should be noted that these countries account for around "40% of global GDP and 10% of the world's population". For many years, Russia was part of this group, although it was expelled in 2014 because it was part of the Ukraine Summit and was seen as violating the sovereignty and integrity of its territory (Fig. 3.1).

Over the years and with the evolution of technologies combined with the increase in productivity in the countries, it is clear that all the countries belonging to this group have seen a substantial increase between the 2 years except Italy, which has a less marked difference. This is because Italy's economy was one of the hardest hit by the 2008 financial crisis, which resulted from a huge number of risky credit loans, and it is also currently the country with the highest public debt compared to other European countries. The country's economic growth was very low at around "0.3% a year". However, Canada's economy, while ranking among the most advanced globally, exhibits a significant reliance on international trade. Approximately one-third of the nation's GDP is attributed to the trade of goods and services, highlighting a substantial dependence on global economic interactions, with its biggest "allies" being the United States of America, China and Japan. Another major world economy is the French economy, which is a mix of private initiative

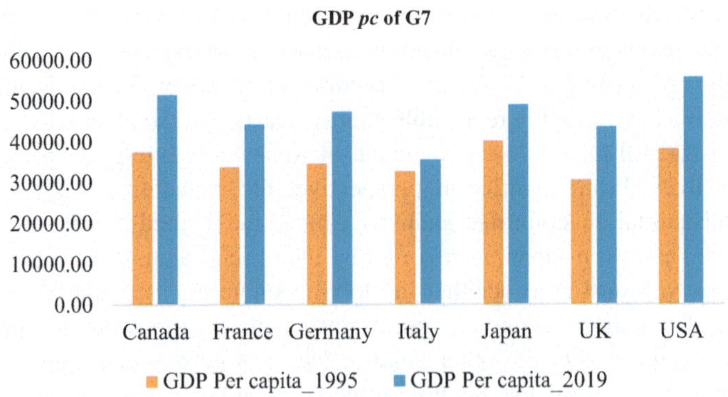

Fig. 3.1 Evolução do PIB per capita dos países constituintes dos G7 no ano de 1995 e 2019 (Fonte: World Bank Data) (https://data.worldbank.org/indicator/NY.GDP.MKTP.CD)

and government involvement. Presently, the service sector plays a paramount role in the country's economy, constituting over 70% of the GDP. Germany's economy is one of the largest in the European Union and the world, as it has high figures when it comes to economic indicators such as GDP, but this is also due to great political support from the country's government, which provides a stable price level and a low and reduced unemployment rate.

Japan stands as one of the world's largest and most developed economies, characterized by a highly educated population and advanced technological infrastructure and diligent labour force combined with a high population density, making it Japan boasts one of the world's largest consumer markets, with its economy ranking second globally, following the United States of America, from 1968 until 2010 when China surpassed it.

The United Kingdom is placed in a very favourable position among the world's biggest and best economies, only being overtaken by Germany in a European context. However, growth in the British economy has slowed since the 2016 referendum to leave the European Union, known as Brexit. The services sector generates employment for almost 81% of the working population, accounting for around 71.3% of the country's GDP.

On the other hand, the banking sector is extraordinarily diligent, as is the tourism sector, which accounts for 10% of GDP.

Finally, the United States has one of the richest and most diversified economies in the world, led by a very productive and efficient service sector and advanced and developed manufacturing. Over the years, the USA has enjoyed relatively robust and strong economic growth, stimulated by tax cuts and tough regulations. This world power constitutes less than 5% of the world's population, yet Americans produce and earn more than 20% of the world's total income.

On the other hand, there is another group of countries called the BRICS, made up of Brazil, Russia, India, China and South Africa. This group is known as the world's emerging economies because the rise of these formerly underdeveloped countries has taken place in just a few years and they have become stronger and more cohesive, and this rise has often been confused with the process of globalization itself. Although there is no clear and succinct definition of what emerging economies are, today they are referred to as such when their economic growth is prosperous and their socio-economic characteristics are different from other peripheral economies.

It was in 2001 that the well-known British economist Jim O'Neil coined the term "BRIC" to refer to the four main so-called emerging countries whose economic potential was quite high, as they showed strong future growth and major improvements in financial indices and indicators. Therefore, these countries decided to adopt this term as a form of informal linkage, just like the G7, even transforming the acronym into "BRICS", with the inclusion of South Africa in 2010, which although it is considered an emerging country has lower figures than those that were already part of the acronym.

One of the main causes for the creation of these countries was the existence of a large quantity of cheap labour, as well as its costs, the very favourable demographics and the abundant natural resources at the time. However, it is important to note that

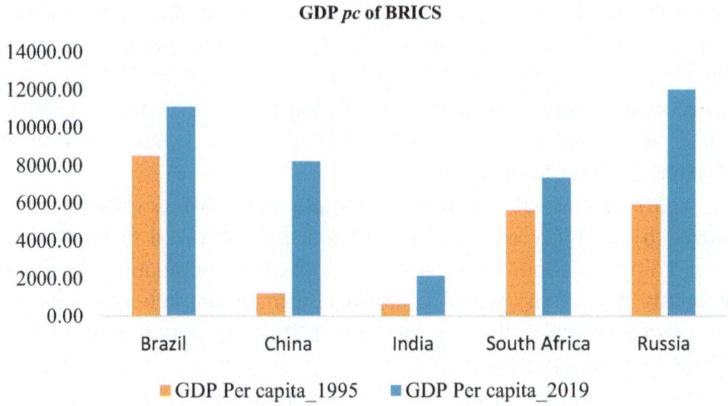

Fig. 3.2 Evolução do PIB per capita dos países constituintes dos BRICS no ano de 1995 e 2019 (Fonte: World Bank Data) (https://data.worldbank.org/indicator/NY.GDP.MKTP.CD)

the Goldman Sachs thesis (economist Jim O'Neill) did not aim for these countries to become a political alliance like, for example, the European Union.

Goldman did, however, say that they would have a high potential for forming a strong economic bloc (Fig. 3.2).

It's significant to highlight that when these informal groups are aggregated, their collective GDP figures are notably high. This economic bloc represents over 40% of the world's population and contributes to over 20% of the global gross domestic product. The economies of these nations have demonstrated consistent growth in economic terms, encompassing not only GDP per capita but also foreign and international trade. China stands out prominently in this regard, while South Africa is distinct for maintaining a consistent participation in world trade over the last two decades.

One of the factors that have made it possible for these countries to increase their growth is the opening up of foreign trade, the increase in the current account balance and foreign exchange reserves since in these parameters the BRICS countries are doing relatively well.

The Brazilian economy is one of the largest in the world and one of the largest in Latin America. However, this country is the best example of the fact that economic growth does not always correspond to economic development, since even though it has high gross domestic product figures, other economic indicators do not necessarily reflect good social indicators.

China has been one of the countries with the highest growth in GDP over the years and is now the world's largest exporter of goods. However, it is important to note that China's economic growth has slowed significantly since 2007, which is believed to have been caused by the US subprime crisis.

India has always had major problems with its economic growth and although it has increased compared to 1995, its figures are extremely low. However, the country is the second most populous in the world and has been growing more significantly, particularly in the industrial sphere, following European decolonization in the second half of the twentieth century. This was due to the abundance of labour at very low costs.

South Africa's economy has seen an increase in its GDP figures over the years, and although great strides have been made in increasing the social welfare of the population, there are still a high number of people living below the poverty line. This is also due to a high unemployment rate per labour force and inequalities among the population, as revealed by the country's Gini Index.

Although Russia's population is highly educated, the country's gross domestic product ranks only 11th in the world, according to figures from the International Monetary Fund (IMF). Russia's economy has grown a lot since Putin took office, although it has had major problems in the face of the 2008 financial crisis and subsequently due to the fall in the value of oil in 2016.

In recent decades, there has been a notable and concerning surge in greenhouse gas emissions, highlighting global warming as a prominent issue of global importance. The escalating levels of CO_2 emissions not only jeopardize the biophysical ecosystem but also significantly influence human society. Hence, it is imperative to comprehend the degree to which these emissions impact a nation's economic growth.

Because carbon emissions are one of the biggest causes of global warming, it became important to include the variable of CO_2 emissions measured in tonnes per capita. This variable is very important because, in addition to the economic growth generated in countries, it is necessary to know how this affects the environment and how much of this growth results in pollution, environmental damage and CO_2 emissions.

However, the increase in population density and trade itself are also factors behind the appearance of these same emissions. Although the theory may show that an increase in economic growth leads to an increase in CO_2 emissions into the environment, there is evidence that may show the opposite, such as the use of less polluting technologies, but this relationship will be confirmed in the practical framework in the second part of this report, making it an uncertain relationship that differs from country to country. However, most of the literature states that there is a negative relationship between CO_2 emissions and economic growth.

3.2 Literature Review

For this purpose, we carried out a brief literature review, where we found that there is a predominance of studies that aim to examine the causal relationship between CO_2 emissions and economic growth for several groups of countries. Thus, on the one hand, some authors defend that economic growth has a harmful and significant negative impact on CO_2 emissions. For example, the works developed by Zubair et al. (2020) and Aslam et al. (2021) support the premise that CO_2 emissions decrease in the face of increases in economic growth. On the other hand, some authors argue that countries with more CO_2 emissions are those that grow more economically. In this alignment, for example, the studies of Wasti and Zaidi (2020) and the study of Llanos et al. (2022) show that an increase in GDP is associated with an increase in CO_2 emissions. Given the importance of renewable energy consumption for the mitigation of greenhouse gases, we consider recent studies that infer that CO_2 emissions decrease as a result of the increase of renewable energy in the energy matrix, such as, for example, the works developed by Li and Haneklaus (2022) and Li and Haneklaus (2021), in which these authors sustain the argument that the option of renewable energy consumption contributes to the decrease of pollutant emissions. The importance of natural resource rents and their contribution to the evolution of pollutant gas emissions has been examined by several studies. The results found in the studies by Agboola et al. (2021) and Nwani and Adams (2021), for example, show that there is a positive and significant link between total natural resource rents and CO_2 emissions. Not least, in the mitigation of greenhouse gases, there is a debate in the literature about the presence of corruption in a given country and its effects on CO_2 emissions. The studies conducted by Wang et al. (2018) and Li and Haneklaus (2021), among others, show that corruption has a positive impact on CO_2 emissions. On the other hand, in the study developed by Ren et al. (2021) they concluded that corruption contributes to the increase of CO_2 emissions in the short term and, in opposition to the same results, evidences a reduction in emissions in the long term. From another perspective, the role of economic, and political uncertainty is a study that is frequently ignored. For example, political uncertainty will influence the level of pollution through increased economic growth. More precisely, the uncertainty of economic policy can influence the business environment and the decision-making of industries and manufacturing companies (Danish et al., 2020). In contrast, Greenhouse Gases (GHG) and carbon emissions are meticulously linked to decisions related to the production of companies and corporations (Jiang et al., 2019). More recently, the study developed by Nakhli et al. (2022) shows, according to its empirical conclusions, that for a complete sample of the Granger causality test, there is bi-directional causality between CO_2 emissions and the uncertainty of economic policy. This variable has been included and we have explored the dynamic association with the drivers to mitigate pollution emissions in 11 selected counties included in the G7 group and BRIC group. We excluded South Africa, as there was no information on the EPU variable.

The sustainability of economic growth should be a cornerstone of governments and political decision-makers, who are responsible for defining and implementing intelligent and sustainable public policies. In this line of reasoning, Azam (2016) advocates that environmental degradation must be controlled without diminishing real growth and the well-being of society so that the desirable level of economic growth can be achieved. Regarding the evidence obtained by Zubair et al. (2020), applied to the Nigerian economy, over the 1980–2018 time horizon, using the ARDL model, the results of the estimations suggest that CO_2 emissions decrease in the face of increased economic growth. Using Fully Modified Ordinary Least Square (FMOLS) estimators, Hdom and Fuinhas (2020) analysed the economies of Brazil using cointegration regression methods: FMOLS (Fully Modified Ordinary Least

Square) and DOLS (Dynamic Ordinary Least Square) and concluded that gross domestic product harms CO_2 emissions. For their part, Zhang et al. (2021) estimated a DEA (Data Envelopment Analysis) model to measure energy and environmental efficiency for selected Central and Western European countries between 2010 and 2014, concluding that effective reforms in the energy sector for countries with lower energy efficiency still face the problem of environmental degradation. Caglar et al. (2022) analysed the economies of the BRICS (Brazil, Russia, India, China and South Africa) between 1990 and 2019 and found that CO_2 emissions decrease in the face of increases in economic growth, according to the ARDL (Auto-regressive distributed lag) model, CO_2 emissions decrease in the face of increases in economic growth. In contrast to previous evidence, Wasti and Zaidi (2020) analysed the relationship between CO_2 emissions and GDP in the context of the Kuwaiti economy, with annual data from 1971 to 2017, concluding that economic growth accelerates CO_2 emissions, using unit root tests such as the Augmented Dicky Fuller and Phillips-Perron methods. Llanos et al. (2022), using unbalanced panel data for 94 countries over the 1971–2018 time horizon, confirm that an increase in GDP is associated with an increase in CO_2 emissions, using a methodological approach in which time series are transformed into a set of treatment intervals, according to the potential policies that each country can adopt. Pejović et al. (2021) explore the relationship between economic growth and CO_2 emissions for the 27 countries of the European Union and the Western Balkans in the period 2008–2018, applying a panel vector autoregressive approach, and find that the higher the growth of a given economy, the higher the CO_2 emissions. As the above shows, in the reference literature, countries' economic growth is directly related to the environment and CO_2 emissions. The environmental Kuznets curve is a theoretical framework in the study of the relationship between CO_2 emissions and economic growth (Sinha & Shahbaz, 2018), representing a measure of the level of environmental degradation. For AlKhars et al. (2022), as the economy grows, cleaner and less polluting technologies are used to compensate for the observed level of environmental degradation. As a result, the country can continue to grow in economic terms, even if this means progressively lower amounts of CO_2 emissions. The study by Alsamara et al. (2018) uses panel data for the 1980–2017 time horizon and contributes to the ongoing debate by validating the Kuznets environmental curve hypothesis, pointing to the verification of a U-inverted configuration to represent the relationship between GDP and CO_2 emissions, using cross-sectional dependence, unit root tests, panel integration, FMOLS (Fully Modified Ordinary Least Square) estimation, PMG (Paymaster General) estimation and dynamic models. Acar et al. (2018) analysed developing countries over the 1970–2016 time horizon using the fixed effects model and the GMM estimator and found that environmental degradation increases as wealth generation in a given country also increases, advocating that cleaner technologies should be introduced ratifying the aforementioned inverted U-shaped curve configuration, Jóźwik et al. (2021) analysed Central European economies in the period 1995–2016, using the econometric technique of cointegration, and concluded that growth is partially related to a reduction in CO_2 emissions, although the latter continues to be a serious problem to combat. Bearing in mind the view of Rashid Gill et al. (2018), economies show behaviour that adequately adheres to the configuration originally proposed for the environmental Kuznets curve, only later to show environmental concerns, which can lead to high costs in the future. For the same authors, in critical terms, each variable that is extended to the conjecture proposed by Kuznets only captures a fraction of the initial problem of the environmental Kuznets curve, and it is considered necessary to include other groups of determining factors that make it possible to operationalize an improvement in environmental quality, namely through sustainable economic growth and improving the quality of both economic policies and institutions.

The ongoing debate on greenhouse gas mitigation is worth highlighting the importance attributed to the perception of corruption in a given country and its effects on CO_2 emissions. Wang et al. (2018), using a partial least squares regression model for a panel of BRIC countries over the period 1996–2015 shows that the moderating role of corruption is crucial in the relationship between economic growth and CO_2 emissions, with the control of corruption helping to slow down the level of CO_2 emissions. Sinha and Shahbaz (2018) argue that corruption increases environmental degradation as well as CO_2 emissions, using cross-sectional dependence tests, unit root tests and cointegration tests for the BRICS and Next-11 countries over the period 1990–2017. Bimonte and Stabile (2019) used panel data regression models for the economies of Italy, concluding that corruption increases CO_2 emissions for the period 1980–2008. Li and Haneklaus (2021) analysed 33 countries in Asia for the period 2000–2015, concluding that corruption also contributes to environmental pollution, using unit root and cointegration tests. For ten countries in Western Central Europe in the period 1990–2019, Simionescu et al. (2022) used the ARDL estimator and concluded that the control of corruption helps to improve the environmental quality of countries. Chen et al. (2022) with parent-level data for the years 2000–2016, carried out heteroscedasticity and endogeneity tests and concluded that countries that control corruption in their economies have lower CO_2 emissions. Also in the context of the Kuznets environmental curve conjecture, Usman (2022) carried out a study applied to the Nigerian economy between 1990 and 2016, using ARDL models, revealing that the more corrupt a country is, the greater its environmental degradation will be. In summary terms, there is no convergence of empirical results, because concerning the relationship between corruption and CO_2 emissions, there is evidence that corruption contributes to an increase in greenhouse gas emissions, but there are also studies that highlight the significant importance of controlling corruption as a mechanism for mitigating the level of environmental degradation. Economic policy uncertainty can influence the level of pollution by increasing economic growth. However, the role of economic policy uncertainty in determining CO_2 emissions is still a relatively unexplored issue. More precisely, economic policy uncertainty can influence the business environment and decision-making processes of different economic activities industries and manufacturing companies (Danish et al., 2020). The same authors analysed and concluded that economic policy uncertainty strengthens the negative effect of energy intensity on CO_2 emissions, using the ARDL model. For Jiang et al. (2019), carbon emissions are linked to business decisions related to the production of companies and economic groups. In the study applied to the United States of America, the same authors, using the parametric Granger causality test, concluded that carbon emissions are affected by the Economic Policy Uncertainty Index when the growth of carbon emissions is in a period of higher or lower growth. More recently, in the study developed by Sarwar et al. (2020) and applied to the Chinese economy in the period 1984–2018, a bidirectional Granger causality relationship was detected between the Economic Policy Uncertainty Index and CO_2 emissions. In the same study, it was found that CO_2 emissions are positively affected by the Economic Policy Uncertainty Index (i.e. the greater the economic policy uncertainty, the greater the level of environmental degradation). Following the hypothesis of the environmental Kuznets curve, the study by Vitenu-Sackey and Acheampong (2022) also shows that high levels of economic policy uncertainty have a positive and significant influence on CO_2 emissions. This evidence was based on an analysis of 18 developed country economies over the period 2005–2018, using three robust long-term estimators, i.e. the two-stage least squares (2SLS) estimator, the generalized method of moments (GMM) and the generalized least squares (GLS) estimator. About the Economic Policy Uncertainty Index, the literature shows that its relationship with CO_2 emissions differs substantially. On the one hand, it can be seen that greenhouse gas emissions are positively affected by the Economic Policy Uncertainty Index, but on the other hand, this index strengthens the negative effect of energy intensity on CO_2 emissions.

Given the importance of renewable energy consumption to mitigate greenhouse gases, we also consider recent studies which suggest that CO_2 emissions will decrease as a result of increased consumption of renewable energy in the energy matrix. Thus, Li and Haneklaus (2021) test the adequacy of the Kuznets environmental curve model for China for the period 1990–2020, using a specification of an ARDL model, concluding that renewable energy consumption reduces CO_2 emissions. Shafiei and Salim (2014) analyse the determinants of CO_2 emissions using the STIRPAT model applied to OECD countries for the period 1980–2001, concluding that the consumption of non-renewable energy contributes to an increase in CO_2 emissions, while the consumption of renewable energy contributes to a decrease in CO_2 emissions. For Middle Eastern and North African countries, Alharthi et al. (2021) found that renewable energy consumption significantly reduces the level of CO_2 emissions over the period 1990–2015, using quantile regression. The study by Li and Haneklaus (2022) applied to the G7 economies, using the ARDL estimator, over the period between 1979 and 2019, suggests that opting for renewable energy consumption contributes to mitigating pollutant emissions. Gill et al. (2018) also maintain that the consumption of renewable energy contributes to reducing CO_2 emissions, which supports the option of the former also being empirically tested in the Kuznets environmental curve conjecture. On the other hand, and in contrast to energy from renewable sources, there are non-renewable sources, for which numerous studies have also been carried out on their relationship with CO_2 emissions. As an example (AlNemer et al., 2023), using Saudi Arabia between 1963 and 2020 and Granger causality tests, concluded that the consumption of non-renewable energies increases the amount of CO_2 emissions. Along the same lines, Karaaslan and

Çamkaya (2022) analysed Turkey between 1980 and 2016 and, using the ARDL method concluded that the higher the consumption of non-renewable energy, the higher the levels of CO_2 emissions. The study carried out by Vo and Vo (2022) for the economies of countries at different levels of development, using a vector autoregressive (VAR) model for the period between 1968 and 2015, suggests that increases in non-renewable energy lead to higher levels of CO_2 emissions. Kang et al. (2019) use a VAR for the period 1965–2015 with a focus on India, concluding that CO_2 emissions increase with higher levels of growth in energy consumption from non-renewable sources.

As far as the relationship between CO_2 emissions and energy consumption is concerned, previous empirical evidence converges on two fundamental points, namely the verification of (1) a negative relationship between renewable energy consumption and CO_2 emissions; and (2) a positive relationship between non-renewable energy consumption and the level of environmental degradation. In conclusion, the main difference between the model we propose to estimate and the many others already existing in the literature, some of which are presented in this section, is the possibility of this model relating institutional and environmental variables in an expansion of the well-known Kuznets Environmental Curve model. In other words, the impact of environmental and energy variables on CO_2 emissions will be analysed and, on the other hand, the impact of institutional variables on CO_2 emissions and economic growth will also be studied in both analyses. This new model combats some of the flaws in the studies presented since new variables significantly impacting CO_2 emissions are added to the well-known Kuznets Environmental Curve.

3.3 Methodology and Data

More data available in a panel data model permits the relaxation of some previously imposed premises and makes it possible to focus on building a more realistic characterization of the countries' problems (Belotti et al., 2013). The panel data SFA technique is more efficient was proposed by Greene (2005) using a time-varying stochastic frontier (SF) normal half-normal model with unit-specific intercepts (Belotti et al., 2013). The random effects assume that the country inefficiency is the same every year on the period observed (Greene, 2005). Using the equation below, the SFA estimate the "true" random effects

$$y_{it} = \alpha_i + \beta'_{it}x + w_{it} + v_{it} \pm u_{it} \tag{3.1}$$

where $\beta'_{it}x$ is an ordinary random effects model, w_{it} is the random country-specific effect, v_{it} is the symmetric and u_{it} is the sided components earlier specified (Greene, 2005). This appears to be a model with a three-part disturbance, but it is a random effects model where the time component does not have a symmetric distribution (Greene, 2005). This random model is not estimated using a maximum likelihood, as is commonly done, it is necessary to integrate a common term from the probability function, and integration can be done through quadrature or simulation (Greene, 2005). In this way, the model can be written equivalently as an SF with a country-specific random term, as follows.

$$y_{it} = (\alpha_i + w_i) + \beta'_{it}x + v_{it} \pm u_{it} \tag{3.2}$$

The "true" fixed model is a standard panel data model, in this approach, α_i is an unobservable individual effect (Xiaoyan Zheng & Heshmati, 2020). The estimation is conducted through a group of ordinary least squares using dummy variables. This model interprets the α_i considering as the country-specific inefficiency based on $\widehat{\alpha}_i^* = max_i\widehat{\alpha}_i - \widehat{\alpha}_i$ or $\widehat{\alpha}_i^* = min_i\widehat{\alpha}_i - \widehat{\alpha}_i$ (Greene, 2005). The higher α_i estimated value is used as a frontier reference (Zheng & Heshmati, 2020). However, this methodology presents certain limitations. Specifically, it does not distinguish individual heterogeneity from inefficiency. Moreover, if the observed period (T) is not sufficiently large, the assumption of time-invariant inefficiency can pose a challenge (Moutinho & Madaleno, 2023). So, to avoid the second problem, this approach is recommended for a large sample, because uses the "brute force" technique, in which the stochastic frontier simply creates dummies variables (Greene, 2005). This model is estimated through the equation below, following (Kumbhakar & Heshmati, 1995):

$$y_{it} = \alpha_i + X'_{it}\beta + v_{it} \pm u_{it} \tag{3.3}$$

Demonstrating the specification of technical inefficiency ($u_{it} = \mu_i +$, where μ_i is the persistent component and τ_{it} the residual component of inefficiency, both are negative) allows for random variation between countries over time (Kumbhakar & Heshmati, 1995). This provides the advantage of being able to test the presence of inefficiency without forcing a time-dependent parametric parameter (Kumbhakar & Heshmati, 1995). Finally, there are four components of the model with determinants of inefficiency; this model does not present the problems of the two previous models (Moutinho & Madaleno, 2023). In this approach, the term of the error is divided into four components (Xiayoan Zheng & Heshmati, 2020). In this chapter, the four components are (1) Hospitals' latent heterogeneity; (2) short-run time-varying transitory inefficiency; (3) persistent or time-invariant inefficiency; and (4) random shocks (Moutinho & Madaleno, 2023). Thus, the model can be written as follows:

$$y_{it} = \alpha_0 + f(x_{it}; \beta) + \mu_i + v_{it} + \eta_i + u_{it} \tag{3.4}$$

where μ_i, v_{it}, η_i and u_{it} involve two-sided individual hospital heterogeneity, two-sided random error components, one-sided time-invariant individual inefficiency, and one-sided time-variant inefficiency (Moutinho & Madaleno, 2023). The signals from η_i to u_{it} are negative in a production model because they represent the inefficiency, showing operational costs below the hospital's ability to produce (Moutinho & Madaleno, 2023).

We follow the basic frontier stochastic SFA model developed by Aigner et al. (1977), with parametric specification involving a logarithmic transformation of the technical efficiency measure which relates the observed output, y_{it}, to the production frontier, $f(x_{it};\beta)$, and adding a random error term, as presented in (1):

$$\ln y_{it} = \ln f(X_{it;}\beta) + v_{it} - u_{it} \tag{3.5}$$

where $y_{it} = a_{it}f(X_{it};\beta)$, with $0 < a_{it} \geq 1$ and $u_{it} = - \ln a_{it} \geq 0$, representing the technical inefficiency. However, we will have proposed a formulation of a stochastic boundary defined from a translog function, given by the following eq. (2):

$$\ln y_{it} = \beta_0 + \sum_{i=1}^{N} \beta_{it} \ln X_{it} + \frac{1}{2} \sum_{i=1}^{N} \sum_{j=1}^{N} \lambda_{itit} \ln X_{it} \ln X_{jt} + v_{it} - u_{it} \tag{3.6}$$

in which the composed error term ε_{it} is given by the difference of a disturbance term called v_{it},and a non-negative variable associated with the technical inefficiency of the country namely u_{it}, which represents inefficiency. As in any conventional regression model that is adopted, the perturbation term v_{it} is responsible for random factors not observed. The symmetric disturbance, v_{it}, is assumed to be due to uncontrollable factors such as weather, making the frontier stochastic. In this study, we estimated the stochastic frontier analysis according to the assumptions connected to inefficiency behaviour, and we used time-invariant inefficiency in the frontier models developed by Pitt and Lee (1981) and Battese and Coelli (1988). In addition, we used the frontier models of variant inefficiency in stochastic frontier models developed by Battese and Coelli (1992), Kumbhakar (1990) and Battese and Coelli (1995). It should also be noted that, in Battese and Coelli (1995), the term "composed error" that measures technical inefficiency includes explanatory variables that make it possible not only to estimate the technical change at the border, as well as estimate the variant technical inefficiency in time. In the case of time-invariant models (Pitt & Lee, 1981) were the first to extend the model presented in Eq. (3.1) to longitudinal data. They proposed maximum likelihood estimation (MLE) of the half-normal SF model, where $u_i \sim N(0, \sigma_u^2)$—This implies that the probability density function of each u_i is a truncated version of a normal random variable with zero mean and a certain variance. σ_u^2—and $v_i \sim N^+(0, \sigma_v^2)$, which says that the error term v_i. These variables are independently and identically distributed as normal random variables with zero means and variances σ_v^2. The generalization of this model to the normal-truncated normal case has been proposed by Battese and Coelli (1988). The time-invariant model is estimated by the employment of the long-run or persistent inefficiency is estimated using the Maximum Likelihood Estimation (MLE) method, and it is determined by the equation presented in the study (2), where $E\,(u_{it}\,|\,v_{it}-)$.

3.4 Empirical Results

According to Table 3.1, the results show different impacts, with 1% or 5% statistical significance levels for the selected explanatory variables on carbon dioxide emissions, so, in particular there, is an inverse relation between the GDP per capita and CO_2 emissions per capita, ceteris paribus, for all the models estimated. This evidence is in line with the results of the studies by Zubair et al. (2020) and Aslam et al. (2021) among others.

However, regarding the nexus between natural resource rents and per capita carbon emissions, there is a negative effect with a significance at the level of 5% only in the model time-varying BC time (92) and in the model for varying inefficiency in BC (88). These results present evidence opposite to the results of the studies by Agboola et al. (2021) and Nwani and Adams (2021).

For the expected validity of the effect between renewable energy consumption and CO_2 emissions per capita, only the results of the time-varying BC model (95) and the time-varying of Kumb (90) confirm the negative relation with a significance of 1%, and these results are corroborated in the studies of Li and Haneklaus (2022) and Li and Haneklaus (2021), among others. In the stochastic boundary estimation in the Kuznets analysis, it should be noted that the corruption perception index has a positive and significant effect on CO_2 emissions per capita in the estimation of time-varying models BC (92) and Kumb (90) and time-invariant model PL (88). This evidence is corroborated in the studies by Wang et al. (2018) and Li and Haneklaus (2021), among others. In the relationship proposed by the analysis of the Kuznets curve, in which we considered the effect of the uncertain economic policy index (EPU) on carbon emissions, only the results of the time-varying of BC (92) show significant statistical evidence at the level of 1% to validate the premise that the EPU effect on CO_2 emissions per capita is negative. At the same time, the estimation results of the time-invariant model of PL (81) have a positive and significant effect on 1% of the EPU on CO_2 emissions per capita. The results of the interactive effects of GDP per capita with all other explanatory variables considered in the proposed EKC show a positive and significant effect on per capita carbon emissions in the stochastic boundary models at 1% or 5% significance for all estimated models. In turn, the interactive effect between the perception index of corruption and the economic, and political uncertainty index, has a negative and significant effect on CO_2 emissions per capita with 1% significance according to the results evidenced only in the estimated time-varying Kumb (90) and time-invariant model of PL (81) frontier model. Whereas, the interactive effect of the corruption perception index on renewable energy consumption shows a negative and significant effect on CO_2 per capita emissions with 1% significance according to the results of the time-varying models BC (92) and Kumb (90) and with the model of time-invariant model of BC (88). In turn, we highlight that the effect between the EPU index, and the consumption of renewable energy, has a negative and significant effect with a 1% significance level, according to the results of the time-varying model of Kumb (90) and the time-invariant model of PL (81). Finally, as the main

Table 3.1 Results of estimation for CO_2 emissions technical inefficiency for G7 + BRIC group countries

Frontier models	Batesse and Coelli (BC 92) Coefficient	Batesse and Coelli (BC 95) Coefficient	Kumbhakar (Kumb 90) Coeffcient	Batesse and Coelli (BC 88) Coefficient	Pitt and Lee (PL 81) Coeffcient
Ln GDP per capita	−7.9560***	−10.0023***	−15.59***	−7.6656***	−15.45***
Ln Rents of resources	−1.9627**	−1.7091	−3.0853	−2.7678***	−1.3113
Ln Corruption	13.1518***	8.8076	18.964***	12.4910***	18.1677
Ln EPU	−3.2160***	2.9445	4.9861	−3.1354	4.8187***
Ln Renewable consumption	2.4779**	−7.4546***	−8.283***	1.5233	−7.9041
Ln GDP per capita quadratic	0.2808**	0.2458**	0.4758***	0.2290**	0.5313***
Ln Rents of resources quadratic	0.0773***	0.0565	0.1050***	0.0995***	0.0490***
Ln Corruption quadratic	−4.2778	−0.3745	−1.461***	−4.0989	−1.5684
Ln EPU quadratic	0.2745***	−0.0571	−0.111***	0.2189***	−0.111***
Ln Renewable energy quadratic	−0.5115***	0.2961**	0.3991***	−0.3885***	−0.251***
Ln GDP pc *Ln Rents of resources	0.2932***	0.2805	0.2881***	0.2324***	0.2481***
Ln GDP pc * Ln Corruption	2.6597***	0.0808	0.2136***	2.5619***	0.1780
Ln GDP pc * Ln EPU	0.1819	0.0319	0.0632***	0.2103**	0.0921***
Ln GDP pc *Ln Renewables	0.5982**	1.3220***	1.7353***	0.5679***	1.4698
Ln Ln Rents of resources* Ln Corruption	−0.1299	−0.2083	0.1383	0.2451**	−0.2411
Ln Ln Rents of resources* Ln EPU	−0.0433	−0.1439	−0.066***	−0.0371	−0.0395**
Ln Ln Rents of resources* Ln Renewable	0.0178	0.1366	0.0039	−0.0323	0.0187
Ln Corruption* Ln EPU	−0.4093	−0.6215	−1.072***	−0.4188	−1.072***
Ln Corruption* Ln Renewable energy	−2.0114***	−1.3119	−2.067***	−1.7668***	−1.4675

(continued)

Table 3.1 (continued)

Frontier models	Batesse and Coelli (BC 92) Coefficient	Batesse and Coelli (BC 95) Coefficient	Kumbhakar (Kumb 90) Coeffcient	Batesse and Coelli (BC 88) Coefficient	Pitt and Lee (PL 81) Coeffcient
Ln EPU * Ln Renewable energy	0.1661**	0.0436	−0.0176**	0.2375***	−0.102***
Constant	33.7965	47.8399	51.3794	29.0489	50.6255
U sigma/sigmau_2		3.1239***		1.3327**	
V sigma/sigmav_2		−14.4975***			
Sigma 2	3.9452			3.7914	
Gamma	0.9670			0.9676	
Lambda (Λ)		6705.96***	5.5561***		9.1026***
Log-likelihood	−136.8002	−311.5530	−112.2280	−135.9952	−35.3657
Wald Chi2 (8)	26649.7***	5.0e+07***	90304***	35607.9***	3.9e+0**

Note: * **, and *** mean statistically significance at 10%, 5%, and 1%, respectively

objective in the analysis is to study the form of an Inverted U Shape, alternating with the U-shaped form Kuznets, we point out that when we consider the influence of GDP per capita on emissions per capita for the group of 11 countries in the sample, an Inverted U Shape form is validated for the five boundary model estimates. However, this Inverted U Shape is valid for the ratio of total resource rents to per capita CO_2 emissions, and the relationship between EPU and CO_2 emissions per capita but only in the estimated time-varying model in boundary model BC (92). Regarding the influence of renewable energy consumption on CO_2 emissions per capita, the Inverted U Shape form is validated according to the results of the time-varying boundary models of BC (95) and Kumb (90). Alternating with the U-shaped form is validated in the relationship between the Corruption Index and the per capita CO_2 emissions with the estimated models of time-varying of BC (92) and Kumb (90). According to the results presented in Table 3.2, the efficiency scores for the Carbon Emissions function are derived from the estimates of inefficiency borders shown in Table 3.1. We can point out that within the G7 Group, Canada has the highest value in terms of technical efficiency score according to the estimation results of three stochastic boundary models with time-varying, whereas Germany has the lowest scoring value according to the estimates in all five frontier models considered.

As for the BRIC group of countries, the position of China with the best position in the technical efficiency scoring is highlighted, while India reveals the worst scoring levels, according to the results of all five estimates carried out.

Table 3.2 Results of scores of EKC efficiency using stochastic frontier models

Countries	Batesse Coelli (BC 92)	Batesse Coelli (BC 95)	Kumbhakar (Kumb 90)	Batesse Coelli (BC 88)	Pitt and Lee (PL 81)
Canada	0.9430	0.8769	0.6120	0.0920	0.5456
France	0.3006	0.3532	0.1988	0.0507	0.1852
Germany	0.0094	0.0029	0.0041	0.0009	0.0031
Italy	0.3765	0.6208	0.2061	0.0657	0.1888
Japan	0.4958	0.5703	0.5947	0.1462	0.5597
United Kingdom	0.1536	0.6164	0.4442	0.0501	0.3440
United States	0.2218	0.7801	0.4306	0.0404	0.2747
Brazil	0.3188	0.7835	0.6722	0.1073	0.7109
Russia	0.1761	0.2085	0.2049	0.3314	0.0901
India	0.0746	0.1125	0.1381	0.1018	0.1303
China	0.4307	0.8928	0.7313	0.8657	0.9549

3.5 Conclusion

In this study, the aim was to examine the behaviour of the environmental Kuznets curve (EKC) under economic policy uncertainty conditions using a stochastic frontier analysis. Evidence was collected from 11 selected countries in the G7 and BRIC groups. On the one hand, the principal results of the stochastic frontier for the EKC demonstrate that the Inverted U Shape form is valid for the relationship between total resource rents and CO_2 emissions per capita, and the relationship between EPU and CO_2 emissions per capita, according to the results of the one frontier varying model. On the other hand, Canada has the highest value in terms of technical efficiency score, whereas Germany has the lowest scoring value according to the estimates in all the frontier models considered. As for the BRIC group of countries, China presents the best position in the technical efficiency scoring, while India reveals the worst scoring levels. In summary, according to our results, it is possible to maintain the argument that policymakers and political leadership within these distinct groups should be aware of climate change consequences. To fulfil carbon neutrality targets, they should develop economic policies under economic policy uncertainty that are in line with clean, greener energy and energy security to sustain cleaner production and mitigate pollutant emissions.

Acknowledgements This research received support from NECE-UBI, the Research Unit in Business Science and Economics, which is sponsored by the Portuguese Foundation for the Development of Science and Technology under project UIDB/04630/2020. The funding for this project was provided through national funds by FCT—Fundação para a Ciência e a Tecnologia. It is part of the project: PRIMA/0008/2020, which received the DOI 10.54499/PRIMA/0008/2020 (https://doi.org/10.54499/PRIMA/0008/2020).

References

Acar, Y., Gürdal, T., & Ekeryllmaz, Ş. (2018). Environmental Kuznets Curve for CO2 emissions: An analysis for developing, Middle East, OECD and OPEC countries. *Environmental and Socio-Economic Studies, 6*(4), 48–58. https://doi.org/10.2478/environ-2018-0027

Agboola, M. O., Bekun, F. V., & Joshua, U. (2021). Pathway to environmental sustainability: Nexus between economic growth, energy consumption, CO2 emission, oil rent and total natural resources rent in Saudi Arabia. *Resources Policy, 74*(December), 102380. https://doi.org/10.1016/j.resourpol.2021.102380

Aigner, D., Lovell, C. A. K., & Schmidt, P. (1977). Formulation and estimation of stochastic frontier production function models. *Journal of Econometrics, 6*(1), 21–37. https://doi.org/10.1016/0304-4076(77)90052-5

Alharthi, M., Dogan, E., & Taskin, D. (2021). Analysis of CO2 emissions and energy consumption by sources in MENA countries: evidence from quantile regressions. *Environmental Science and Pollution Research, 28*(29), 38901–38908. https://doi.org/10.1007/s11356-021-13356-0

Ali, E. B., Opoku-Mensah, E., Ofori, E. K., & Agbozo, E. (2023). Load capacity factor and carbon emissions: Assessing environmental quality among MINT nations through technology, debt, and green energy. *Journal of Cleaner Production, 428*, 139282. https://doi.org/10.1016/j.jclepro.2023.139282

AlKhars, M. A., Alwahaishi, S., Fallatah, M. R., & Kayal, A. (2022). A literature review of the Environmental Kuznets Curve in GCC for 2010–2020. *Environmental and Sustainability Indicators, 14*(November), 100181. https://doi.org/10.1016/j.indic.2022.100181

AlNemer, H. A., Hkiri, B., & Tissaoui, K. (2023). Dynamic impact of renewable and non-renewable energy consumption on CO2 emission and economic growth in Saudi Arabia: Fresh evidence from wavelet coherence analysis. *Renewable Energy, 209*(April), 340–356. https://doi.org/10.1016/j.renene.2023.03.084

Alsamara, M., Mrabet, Z., Saleh, A. S., & Anwar, S. (2018). The environmental Kuznets curve relationship: a case study of the Gulf Cooperation Council region. *Environmental Science and Pollution Research, 25*(33), 33183–33195. https://doi.org/10.1007/s11356-018-3161-1

Aslam, B., Hu, J., Shahab, S., Ahmad, A., Saleem, M., Shah, S. S. A., Javed, M. S., Aslam, M. K., Hussain, S., & Hassan, M. (2021). The nexus of industrialization, GDP per capita and CO2 emission in China. *Environmental Technology and Innovation, 23*, 101674. https://doi.org/10.1016/j.eti.2021.101674

Azam, M. (2016). Does environmental degradation shackle economic growth? A panel data investigation on 11 Asian countries. *Renewable and Sustainable Energy Reviews, 65*, 175–182. https://doi.org/10.1016/j.rser.2016.06.087

Baker, S. R., Bloom, N., & Davis, S. J. (2016). Measuring economic policy uncertainty. *The Quarterly Journal of Economics, President and Fellows of Harvard College, 131*(4), 1593–1636.

Battese, G. E., & Coelli, T. J. (1988). Prediction of firm-level technical efficiencies with a generalized frontier production function and panel data. *Journal of Econometrics, 38*(3), 387–399. https://doi.org/10.1016/0304-4076(88)90053-X

Battese, G. E., & Coelli, T. J. (1992). Frontier production functions, technical efficiency and panel data: With application to paddy farmers in India. *Journal of Productivity Analysis, 3*(1–2), 153–169. https://doi.org/10.1007/BF00158774

Battese, G., & Coelli, T. (1995). A model for technical inefficiency effects in a stochastic frontier production function for panel data. *Journal of Econometrics, 20*, 325–332.

Belotti, F., Daidone, S., Ilardi, G., & Atella, V. (2013). Stochastic frontier analysis using Stata. *Stata Journal, 13*(4), 719–758. https://doi.org/10.1177/1536867x1301300404

Bimonte, S., & Stabile, A. (2019). The effect of growth and corruption on soil sealing in Italy: A regional environmental Kuznets Curve Analysis. *Environmental and Resource Economics, 74*(4), 1497–1518. https://doi.org/10.1007/s10640-019-00376-1

Caglar, A. E., Zafar, M. W., Bekun, F. V., & Mert, M. (2022). Determinants of CO2 emissions in the BRICS economies: The role of partnerships investment in energy and economic complexity. *Sustainable Energy Technologies and Assessments, 51*(August), 101907. https://doi.org/10.1016/j.seta.2021.101907

Chen, Z., Hao, X., & Zhou, M. (2022). Does institutional quality affect air pollution? *Environmental Science and Pollution Research, 29*(19), 28317–28338. https://doi.org/10.1007/s11356-021-17934-0

Danish, Ulucak, R., & Khan, S. U. D. (2020). Relationship between energy intensity and CO2 emissions: Does economic policy matter? *Sustainable Development, 28*(5), 1457–1464. https://doi.org/10.1002/sd.2098

Gill, A. R., Viswanathan, K. K., & Hassan, S. (2018). A test of environmental Kuznets curve (EKC) for carbon emission and potential of renewable energy to reduce green house gases (GHG) in Malaysia. *Environment, Development and Sustainability, 20*(3), 1103–1114. https://doi.org/10.1007/s10668-017-9929-5

Greene, W. (2005). Reconsidering heterogeneity in panel data estimators of the stochastic frontier model. *Journal of Econometrics, 126*(2), 269–303. https://doi.org/10.1016/j.jeconom.2004.05.003

Hdom, H. A. D., & Fuinhas, J. A. (2020). Energy production and trade openness: Assessing economic growth, CO2 emissions and the applicability of the cointegration analysis. *Energy Strategy Reviews, 30*, 100488. https://doi.org/10.1016/j.esr.2020.100488

Jiang, Y., Zhou, Z., & Liu, C. (2019). Does economic policy uncertainty matter for carbon emission? Evidence from US sector level data. *Environmental Science and Pollution Research, 26*(24), 24380–24394. https://doi.org/10.1007/s11356-019-05627-8

Jóźwik, B., Gavryshkiv, A. V., Kyophilavong, P., & Gruszecki, L. E. (2021). Revisiting the environmental Kuznets curve hypothesis: A case of Central Europe. *Energies, 14*(12), 3415. https://doi.org/10.3390/en14123415

Kang, S. H., Islam, F., & Kumar Tiwari, A. (2019). The dynamic relationships among CO2 emissions, renewable and non-renewable energy sources, and economic growth in India: Evidence from time-varying Bayesian VAR model. *Structural Change and Economic Dynamics, 50*, 90–101. https://doi.org/10.1016/j.strueco.2019.05.006

Karaaslan, A., & Çamkaya, S. (2022). The relationship between CO2 emissions, economic growth, health expenditure, and renewable and non-renewable energy consumption: Empirical evidence from Turkey. *Renewable Energy, 190*, 457–466. https://doi.org/10.1016/j.renene.2022.03.139

Kumbhakar, S. C. (1990). Production frontiers, panel data, and time-varying technical inefficiency. *Journal of Econometrics, 46*(1–2), 201–211. https://doi.org/10.1016/0304-4076(90)90055-X

Kumbhakar, S. C., & Heshmati, A. (1995). Efficiency measurement in swedish dairy farms: An application of rotating panel data, 1976–88. *American Journal of Agricultural Economics, 77*(3), 660–674. https://doi.org/10.2307/1243233

Li, B., & Haneklaus, N. (2021). The role of renewable energy, fossil fuel consumption, urbanization and economic growth on CO2 emissions in China. *Energy Reports, 7*, 783–791. https://doi.org/10.1016/j.egyr.2021.09.194

Li, B., & Haneklaus, N. (2022). Reducing CO2 emissions in G7 countries: The role of clean energy consumption, trade openness and urbanization. *Energy Reports, 8*, 704–713. https://doi.org/10.1016/j.egyr.2022.01.238

Llanos, C., Kristjanpoller, W., Michell, K., & Minutolo, M. C. (2022). Causal treatment effects in time series: CO2 emissions and energy consumption effect on GDP. *Energy, 249*, 123625. https://doi.org/10.1016/j.energy.2022.123625

Mehmood, U. (2022). Environmental degradation and financial development: do institutional quality and human capital make a difference in G11 nations? *Environmental Science and Pollution Research, 29*(25), 38017–38025. https://doi.org/10.1007/s11356-022-18825-8

Moutinho, V., & Madaleno, M. (2023). Assessing eco-efficiency in Asian and African countries using stochastic frontier analysis. *Energies, 14*, 251–254. https://doi.org/10.4337/9781800377486.stochastic.frontier.analysis

Nakhli, M. S., Shahbaz, M., Ben Jebli, M., & Wang, S. (2022). Nexus between economic policy uncertainty, renewable & non-renewable energy and carbon emissions: Contextual evidence in carbon neutrality dream of USA. *Renewable Energy, 185*, 75–85. https://doi.org/10.1016/j.renene.2021.12.046

Nwani, C., & Adams, S. (2021). Environmental cost of natural resource rents based on production and consumption inventories of carbon emissions: Assessing the role of institutional quality. *Resources Policy, 74*(December), 102282. https://doi.org/10.1016/j.resourpol.2021.102282

Pejović, B., Karadžić, V., Dragašević, Z., & Backović, T. (2021). Economic growth, energy consumption and CO2 emissions in the countries of the European Union and the Western Balkans. *Energy Reports, 7*, 2775–2783. https://doi.org/10.1016/j.egyr.2021.05.011

Pitt, M. M., & Lee, L. F. (1981). The measurement and sources of technical inefficiency in the Indonesian weaving industry. *Journal of Development Economics, 9*(1), 43–64. https://doi.org/10.1016/0304-3878(81)90004-3

Rashid Gill, A., Viswanathan, K. K., & Hassan, S. (2018). The Environmental Kuznets Curve (EKC) and the environmental problem of the day. *Renewable and Sustainable Energy Reviews, 81*(June 2017), 1636–1642. https://doi.org/10.1016/j.rser.2017.05.247

Ren, Y. S., Ma, C. Q., Apergis, N., & Sharp, B. (2021). Responses of carbon emissions to corruption across Chinese provinces. *Energy Economics, 98*, 105241. https://doi.org/10.1016/j.eneco.2021.105241

Sachan, A., Sahu, U. K., Pradhan, A. K., & Thomas, R. (2023). Examining the drivers of renewable energy consumption: Evidence from BRICS nations. *Renewable Energy, 202*, 1402–1411. https://doi.org/10.1016/j.renene.2022.11.080

Sarwar, B., Ming, X., & Husnain, M. (2020). Economic policy uncertainty and dividend sustainability: new insight from emerging equity market of China. *Economic Research-Ekonomska Istraživanja, 33*(1), 204–223. https://doi.org/10.1080/1331677X.2019.1708769

Shafiei, S., & Salim, R. A. (2014). Non-renewable and renewable energy consumption and CO2 emissions in OECD countries: A comparative analysis. *Energy Policy, 66*, 547–556. https://doi.org/10.1016/j.enpol.2013.10.064

Simionescu, M., Neagu, O., & Gavurova, B. (2022). The role of quality of governance in reducing pollution in Romania: An ARDL and Nonparametric Bayesian Approach. *Frontiers in Environmental Science, 10*(April), 1–15. https://doi.org/10.3389/fenvs.2022.892243

Sinha, A., & Shahbaz, M. (2018). Estimation of Environmental Kuznets Curve for CO2 emission: Role of renewable energy generation in India. *Renewable Energy, 119*, 703–711. https://doi.org/10.1016/j.renene.2017.12.058

Uberti, L. J. (2021). Corruption and growth: Historical evidence, 1790–2010. *Journal of Comparative Economics, 50*(2), 321–349. https://doi.org/10.1016/j.jce.2021.10.002

Usman, O. (2022). Modelling the economic and social issues related to environmental quality in Nigeria: The role of economic growth and internal conflict. *Environmental Science and Pollution Research, 29*(26), 39209–39227. https://doi.org/10.1007/s11356-021-18157-z

Vitenu-Sackey, P. A., & Acheampong, T. (2022). Impact of economic policy uncertainty, energy intensity, technological innovation and R&D on CO2 emissions: Evidence from a panel of 18 developed economies. *Environmental Science and Pollution Research, 29*(58), 87426–87445. https://doi.org/10.1007/s11356-022-21729-2

Vo, D. H., & Vo, L. H. (2022). International volatility transmission among income, CO2 emission, non-renewable and renewable energy consumption: Which causes which and when? *Energy Reports, 8*, 10061–10071. https://doi.org/10.1016/j.egyr.2022.07.168

Wang, Z., Danish, Zhang, B., & Wang, B. (2018). The moderating role of corruption between economic growth and CO2 emissions: Evidence from BRICS economies. *Energy, 148*, 506–513. https://doi.org/10.1016/j.energy.2018.01.167

Wasti, S. K. A., & Zaidi, S. W. (2020). An empirical investigation between CO2 emission, energy consumption, trade liberalization and economic growth: A case of Kuwait. *Journal of Building Engineering, 28*(July), 101104. https://doi.org/10.1016/j.jobe.2019.101104

Zeng, Q., & Yue, X. (2022). Re-evaluating the asymmetric economic policy uncertainty, conventional energy, and renewable energy consumption nexus for BRICS. *Environmental Science and Pollution Research*, 1–10.

Zhang, J., Patwary, A. K., Sun, H., Raza, M., Taghizadeh-Hesary, F., & Iram, R. (2021). Measuring energy and environmental efficiency interactions towards CO2 emissions reduction without slowing economic growth in central and western Europe. *Journal of Environmental Management, 279*(May 2020), 111704. https://doi.org/10.1016/j.jenvman.2020.111704

Zheng, X., & Heshmati, A. (2020). An analysis of energy use efficiency in China by applying stochastic frontier panel data models. *Energies, 13*(8), 1–17. https://doi.org/10.3390/en13081892

Zubair, A. O., Abdul Samad, A. R., & Dankumo, A. M. (2020). Does gross domestic income, trade integration, FDI inflows, GDP, and capital reduces CO2 emissions? An empirical evidence from Nigeria. *Current Research in Environmental Sustainability, 2*, 100009. https://doi.org/10.1016/j.crsust.2020.100009

Chapter 4
Carbon Intensity of Products Under the EU CBAM: Assessment of Russia's Competitiveness Based on Inter-Country Analysis

Andrey Yu. Kolpakov and Timur Z. Maimekov

Abstract The implementation of the EU's Carbon Border Adjustment Mechanism (CBAM) expands the concept of competitiveness of products in cross-country trade by adding a carbon footprint metric. An assessment of the carbon intensity of products, to which the CBAM will act (hereinafter CBAM products), is possible on the basis of IEA data and national inventories of anthropogenic emissions. The analysis shows that most Russian goods are characterized by a carbon footprint that is not higher than the weighted average level on the market. The only exceptions are building materials: cement, lime, glass. However, Russia's export of these products to the EU is negligible. Thus, Russia has a worthy position in foreign markets, and the use of only CBAM does not affect the economic attractiveness of Russian goods in the EU.

4.1 Introduction: CBAM and Carbon Footprint as a New Competitive Factor

Several years have passed since the announcement of the EU CBAM, and the moment when this mechanism will work and make the carbon footprint of goods imported into the EU to be paid is getting closer.

The EU explains the need to implement CBAM with its concern regarding the likelihood of transferring European business to the territory of carbon-loyal countries in order to optimize costs (the so-called "carbon leakage"), as well as due to the deterioration in the competitiveness of European goods compared to imported

A. Y. Kolpakov (✉)
Institute of Economic Forecasting of the Russian Academy of Sciences, Moscow, Russia
e-mail: kolpakov@ecfor.ru

T. Z. Maimekov
Moscow Institute of Physics and Technology, Dolgoprudny, Russia
e-mail: maymekov.tz@phystech.edu

© The Author(s), under exclusive license to Springer Nature Switzerland AG 2024

T. Devezas et al. (eds.), *Global Energy Transition and Sustainable Development Challenges, Vol. 1*, World-Systems Evolution and Global Futures,
https://doi.org/10.1007/978-3-031-67583-6_4

analogues. In addition, CBAM can become a tool for the EU to compensate the huge costs of technological modernization of its economy, regardless of whether such goals are declared at the official level or not.

The “design” parameters of CBAM assumed a wide coverage of products and considered the possibility of taking into account not only direct (scope 1), but also indirect (scope 2) emissions of greenhouse gases generated during production. However, in July 2021, the mechanism parameters were announced (European Commission, 2021), which were assessed as quite balanced and cautious (Marcu et al., 2021), although they can be significantly revised in the future. Here are the key characteristics of CBAM.

- CBAM covers only direct (scope 1) emissions of the following aggregated groups of goods: iron and steel, mineral fertilizers, cement, electricity, aluminum, hydrogen. Selected product categories are carbon-intensive and are at the highest risk of carbon leakage.
- CBAM is applied to goods imported from all countries outside the EU, unless the country is already participating/embedded in the EU Emissions Trading System (EU ETS). Such countries are for instance Iceland, Liechtenstein, Norway, Switzerland.
- CBAM would start operating in 2023. However, in 2023–2026 there would be no payments. Instead, importing organizations must report to the EU on imported CBAM goods on a quarterly basis. The reports include data on: (a) the quantity of imported goods; (b) the amount of associated emissions; (c) the amount of carbon tax paid in the country of good origin (if any).
- The first payments under CBAM will start from 2026. The payers are importing organizations that must buy certificates, submit annual declarations, enter the register of product importers, etc. The price of the CBAM certificate is equivalent to the price of the EU ETS.
- CBAM is reduced by the amount of a similar payment, which was explicitly paid in the country of the imported good origin.
- It is important to note that the EU uses a mechanism for distributing free allocations, i.e., emission quotas. Until now, for almost all CBAM goods the volume of free allocations exceeded their carbon footprint (CENEF-XXI, 2021). The only exceptions are ammonia and aluminum, whose producers were forced to pay for about 15% of emissions not covered by free quotas. For CBAM goods, a similar proportional exemption is provided. However, CBAM assumes a gradual reduction in the amount of free allocations in 2026–2034 according to the following schedule: 2026—2.5%, 2027—5%, 2028—10%, 2029—22.5%, 2030—48.5%, 2031—61%, 2032—73.5%, 2033—86%, 2034—100%. During this period, CBAM will only apply to the share of emissions not covered by free allocations. The schedule for phasing out free allocations can be adjusted by the European Commission if necessary (Council of the European Union, 2022).

The period of waiting for the first CBAM payments is characterized by several processes.

On the one hand, European society predominantly supports the implementation of CBAM (Kuehner et al., 2022) since this mechanism will increase the competitiveness of domestic businesses and help (at least not prevent) solving the carbon leakage problem. After CBAM put into action, the EU can increase output and replace part of imports with its own products (Zhong & Pei, 2022). European regulators left the discussion on CBAM compliance with WTO rules behind (Bellora & Fontagné, 2022; Lim et al., 2021) and consider if it's possible and reasonable to tighten CBAM parameters right now. For instance, at the beginning of 2023, the European Commission agreed to shorten the phase-out period for free allocations by 1 year (initially, it was supposed to last not until 2034, but until 2035). In addition, issues of expanding the range of products are being discussed (among the most likely options for expanding are glass, ceramics, cellulose, paper, cardboard, acids, and bulk organic chemicals) and emission coverage (in order to take into account scope 2 emissions).

The arguments in favor of tightening CBAM parameters are obvious: the maximization of positive impacts for European business and the budget. Moreover, CBAM can have significant indirect effect, forcing many countries to reduce the carbon intensity of their economies and address climate change (Magacho et al., 2022). But the limitations are clear: in addition to foreign policy aspects, an aggressive CBAM option will also force suppliers to leave Europe for other markets, thereby causing a shortage of goods in the EU (Böning et al., 2023).

On the other hand, suppliers are carefully analyzing the impact of CBAM on the future physical and financial performance of trade with the EU. The analysis covers both countries closest to the EU and suppliers from the most distant regions (Burke et al., 2021; Chepeliev, 2021; Kaitila et al., 2022; Schott & Hogan, 2022; Takeda & Arimura, 2023). Experts agree that CBAM impact is likely to be insignificant on the scale of the entire economy, but the mechanism is capable of seriously reformatting the markets of particular goods. China, India, and Russia can be characterized by the largest losses of export revenues (Zhong & Pei, 2022). However, the developing countries of Africa and the Middle East, as well as Ukraine, will experience the most significant damage to their wealth in relative (percentage) terms (Chepeliev, 2021; Zimmer & Holzhausen, 2020). Fairly obvious conclusions are done that reducing the negative impact of CBAM is possible through the development of national systems of economic regulation of greenhouse gas emissions, as well as through stimulating industry to modernize production facilities, apply low-emission technologies, increase energy efficiency and energy saving (Qi et al., 2022; Siy et al., 2023).

Research work is also carried out in Russia, as the EU is an important market for a wide range of Russian goods. In 2021, Russia delivered CBAM products of almost \$15 billion worth to the EU (Table 4.1), of which 54% are ferrous metals, 23% are fertilizers, 17% are aluminum, and 6% are electricity. The supply of cement is insignificant on a general scale.

The range of estimates for the future payments under CBAM for Russian goods is 1–2 billion euros per year (IPEM, 2021; Marcu, 2021) (after the complete elimination of free allocations, i.e., after 2035). At the same time, the main concerns regarding CBAM are not only direct payments (which can turn into financial losses

Table 4.1 Russia's exports of CBAM products to the EU in 2021

	Thousand tons	TWh	Million USD
Electricity		14.0	949
Ferrous metals and products	11,530		8005
Aluminum and products	1152		2498
Fertilizers	9331		3365
Cement	49		3
Total			14,820

Source: Federal Customs Service of Russia

for suppliers), but also a possible loss of market niches due to the inability to compete in terms of the carbon footprint of products.

This study is dedicated to a comparative analysis of the carbon intensity of products manufactured in Russia and other countries in order to understand how big the risk is that Russian products will be forced out of the European market due to their carbon footprint.

4.2 Approach to Estimate Carbon Intensity of CBAM Products

The carbon intensity of goods produced in different countries is calculated on the basis of open data.

The assessment of the carbon intensity of electricity production was carried out using the IEA database, namely energy balances and data on energy emissions of greenhouse gases (GHG). These data are generated for each country and published on the IEA website.[1]

The peculiarity of the estimates for the electric power industry is that in several countries Combined heat and power (CHP) plants play a significant role, and they generate electricity and heat simultaneously. At the same time, the volume of fuel consumption in the IEA energy balances is not divided into the needs of electricity and heat production separately. Therefore, in the calculations, the heat generated at the CHP plants is added to the generation of electricity in comparable units of measurement.

The functional record of the calculation algorithm is as follows:

$$GHG_{Ei} = ((C_{PPi} + C_{CHPi}) * c_i + (G_{PPi} + G_{CHPi}) * g_i + (L_{PPi} + L_{CHPi}) * l_i)/(E_i + H_{CHPi}) \quad (4.1)$$

[1] https://www.iea.org/data-and-statistics

$$c_i = \frac{GHGC_i}{(TESC_i - NEC_i)}, g_i = \frac{GHGG_i}{(TESG_i - NEG_i)}, l_i = \frac{GHGL_i}{(TESL_i - NEL_i)} \quad (4.2)$$

where:

GHG_{Ei}—specific GHG emissions from electricity production in a country i;
C_{PPi}, G_{PPi}, L_{PPi}—consumption of coal, natural gas, and liquid fuels, respectively, at power plants in a country i;
C_{CHPi}, G_{CHPi}, L_{CHPi}—consumption of coal, natural gas, and liquid fuels, respectively, at CHP plants in a country i;
c_i, g_i, l_i—specific GHG emissions from the combustion of coal, natural gas, and liquid fuels, respectively, in a country i;
E_i—electricity production in a country i;
H_{CHPi}—heat production at CHP plants in a country i;
$GHGC_i$, $GHGG_i$, $GHGL_i$—GHG emissions from the combustion of coal, natural gas, and liquid fuels, respectively, in a country i;
$TESC_i$, $TESG_i$, $TESL_i$—total energy supply (primary consumption) of coal, natural gas, and liquid fuels, respectively, in a country i;
NEC_i, NEG_i, NEL_i—non-energy use of coal, natural gas, and liquid fuels, respectively, in a country i.

Specific GHG emissions from the production of all other types of CBAM products is calculated as the ratio of annual associated GHG emissions and their annual production volume, and the initial data are taken from reports[2] on the national inventory of anthropogenic GHG emissions from sources and removals by sinks. For such reports, the IPCC recommended a single emission reporting form for all countries, which allows for a comprehensive analysis for reporting countries. At the same time, this reporting is presented by a limited number of countries (about 50 countries: all developed and several developing countries), which narrows the scope of the analysis. For estimates, we used data from the inventory forms "Table 1S1. Sectoral report for energy", "Table 2(I)S1 sectoral report for industrial processes and product use".

The calculation algorithm is as follows:

[2] https://unfccc.int/process-and-meetings/transparency-and-reporting/reporting-and-review-under-the-convention/greenhouse-gas-inventories-annex-i-parties/national-inventory-submissions-2023

$$GHG_{pi} = \frac{\sum_{j=1}^{N}\left(FE_{pi}^{CO2} + 25FE_{pi}^{CH4} + 298FE_{pi}^{N2O}\right)}{P_{pi}} + \frac{\sum_{j=1}^{N}\left(IND_{pi}^{CO2} + 25IND_{pi}^{CH4} + 298IND_{pi}^{N2O} + IND_{pi}^{F}\right)}{P_{pi}} \quad (4.3)$$

where:

GHG_{pi}—specific greenhouse gas emissions from the production of a product p in a country i;

FE_{pi}^{CO2}, FE_{pi}^{CH4}, FE_{pi}^{N2O}—emissions of CO_2, CH_4, and N_2O, respectively, as a result of fuel combustion in the production of a product p in a country i. To convert all types of emissions into unified measurement units (CO_2-eq.) coefficients 25 and 298 are applied for CH_4 and N_2O, respectively;

IND_{pi}^{CO2}, IND_{pi}^{CH4}, IND_{pi}^{N2O}, IND_{pi}^{F}—industrial emissions of CO_2, CH_4, N_2O, and F-gases, respectively, as a result of the production of a product p in a country i;

P_{pi}—amount of the production of a product p in a country i.

In the national inventories, industrial emissions are detailed for each product (e.g., steel, iron, ferroalloys, etc.), and energy emissions are presented for industries (e.g., metallurgy). As part of the calculation, a hypothesis was made that the product structure of industrial and energy emissions in each considered industry is the same. This allows us to break down energy emissions into particular products.

4.3 Results: Inter-Country Analysis of Carbon Intensity of CBAM Products

The results obtained are shown in Figs. 4.1, 4.2, 4.3, 4.4 and 4.5.

The carbon footprint of Russian electricity is one-third lower than the global weighted average level. According to this indicator, Russia is second only to the EU countries (but far from all: in Poland, Greece, the Netherlands, the Czech Republic, Bulgaria, Germany, electricity turned out to be significantly more carbon-intensive than in Russia), as well as several countries with a high share of hydroelectric power plants (Norway, Brazil, Canada).

At the same time, it is important to note that all other large countries that are trading partners of the EU produce electricity with higher specific emissions. This means not only that the export of electricity to the EU will be difficult (significant trade in electricity is, in principle, a rare occurrence), but also high values of indirect emissions from the supply of products to the EU. In other words, if the EU CBAM covers scope 2 emissions, this will affect Russian products to a lesser extent.

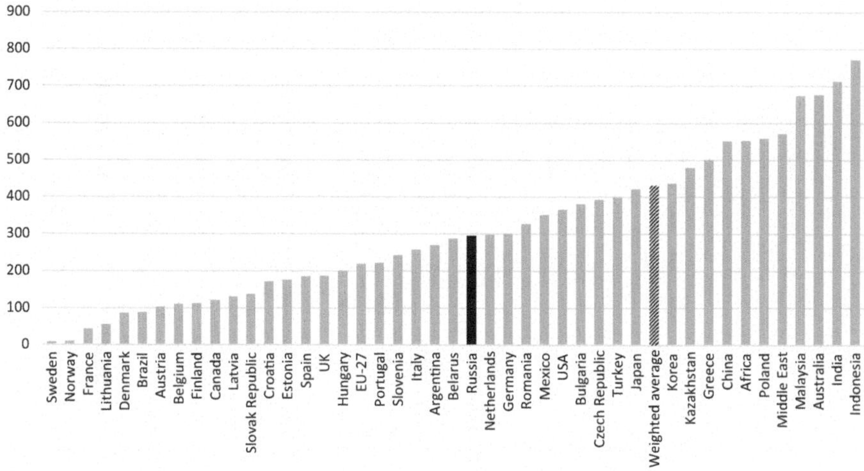

Fig. 4.1 Average carbon intensity of electricity in 2019, g CO_2-eq./kWh

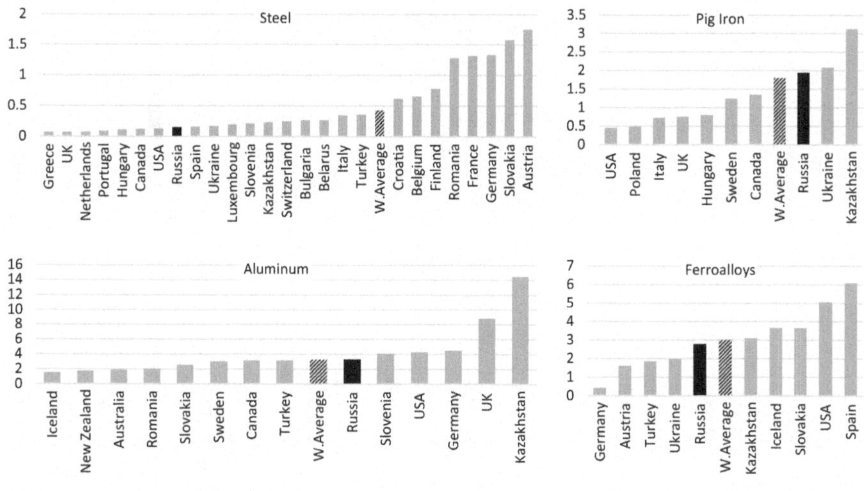

Fig. 4.2 Average carbon intensity of metals in 2019, t CO_2-eq./t

In terms of metallurgical and chemical products, Russia is characterized by a carbon intensity that is either lower or close to the market-weighted average value. In several positions, Russian products outperform such developed countries as the USA, Poland, Germany, as well as developing Kazakhstan, Ukraine, Belarus, and Turkey. In some cases, this is due to the high role of coal in meeting the demand for energy in production, in others—the level of technology used.

It can be expected that Russia will be able to compete in foreign markets in these commodity groups. At least the CBAM mechanism will have an impact primarily on other countries, while Russian manufacturers have a certain margin of safety.

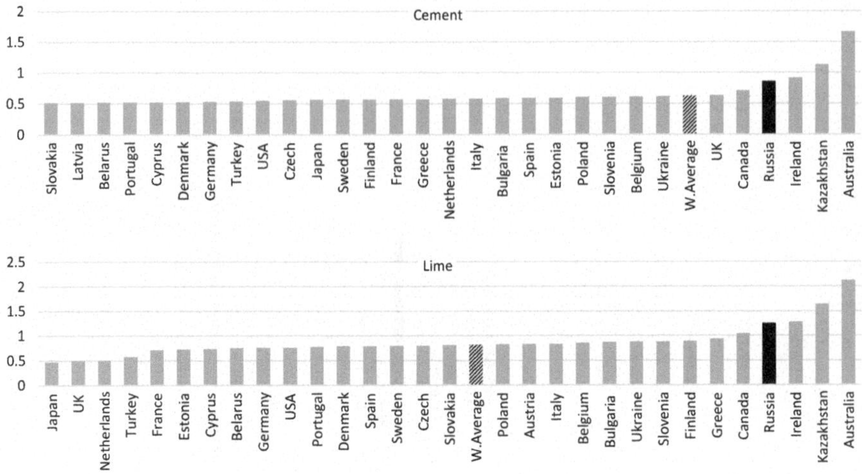

Fig. 4.3 Average carbon intensity of cement and lime in 2019, t CO_2-eq./t

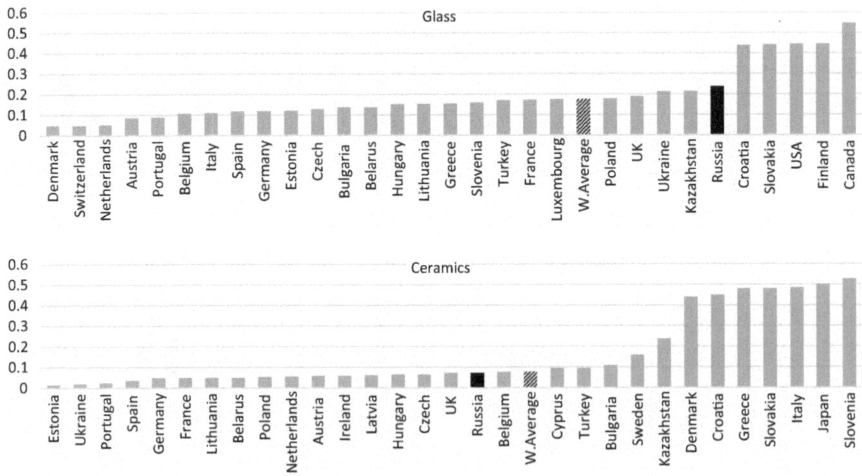

Fig. 4.4 Average carbon intensity of glass and ceramics in 2019, t CO_2-eq./t

Unfortunately, Russia has a high carbon footprint in the building materials industry. For cement and lime, only Kazakhstan and Australia have consistently worse performance (due to the high energy intensity of the processes and the use of coal); on glass—in the USA, Canada, Finland.

However, it should be borne in mind that, of the listed building materials, only cement falls under the CBAM. The export of cement from Russia to the EU ETS countries amounted to only 49 thousand tons and $ 3 million in 2021. In other words, the potential damage from the cessation of such supplies rated as extremely low.

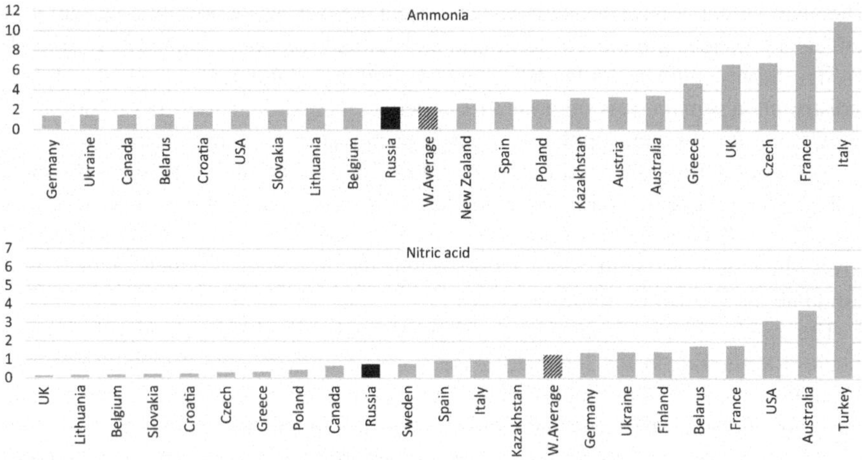

Fig. 4.5 Average carbon intensity of chemicals in 2019, t CO_2-eq./t

Table 4.2 Percentage of supply that is more carbon intensive than Russia

Product	Supply share
Steel	52%
Pig iron	19%
Ferroalloys	50%
Aluminum	17%
Cement	4%
Lime	3%
Glass	6%
Ceramics	29%
Ammonia	15%
Nitric acid	49%

Source: Authors

4.4 Conclusions: Russia's Competitiveness and Sanctions

The data obtained show that the main volume of Russian products that fall under the CBAM is competitive in the European market not only in terms of cost characteristics (which is indicated by significant volumes of supplies), but also in terms of carbon footprint.

In terms of carbon intensity, Russia does not exceed the weighted average among all suppliers. This means that emission control measures applied on a non-discriminatory basis (the same to all) will hit other suppliers first. For most products, at least 15–50% of the supply volume is characterized by a higher carbon intensity compared to Russian producers (Table 4.2).

However, in 2022, developed countries, including Europe, introduced strict foreign trade restrictions against Russia, including in relation to CBAM products.

The main blow fell on ferrous metals: at present, deliveries are prohibited for all iron and steel products. Supplies of fertilizers and cement were also banned or restricted. In significant volumes, only deliveries of aluminum and ferrous metals of low redistribution remain. It is obvious that under such conditions the risks of introducing CBAM for Russian suppliers are disproportionately less significant.

At the same time, if sanctions are eased or lifted in the future, Russian products will take their niche in the European market. Otherwise, a situation is likely when, as CBAM is rolled out and free quotas are cancelled, the EU countries will be forced to cover their demand with more carbon-intensive imported products, which will stimulate the growth of domestic prices.

The results obtained show the competitiveness of different countries in the medium term. But in the long term, the scale of implemented projects to reduce the carbon intensity of products will be decisive. In modern conditions, such projects enjoy significant preferences that increase their investment attractiveness, especially in the advanced economies and countries importing hydrocarbons (e.g., China) (IEA, Imperial College Business School, 2021). Without additional stimulus measures, developing countries risk becoming marginal suppliers in many markets that will be regulated through CBAM-type mechanisms.

References

Bellora, C., & Fontagné, L. (2022). *EU in search of a WTO-compatible carbon border adjustment mechanism*. CEPII Working Paper. http://www.cepii.fr/PDF_PUB/wp/2022/wp2022-01.pdf

Böning, J., Di Nino, V., & Folger, T. (2023). *Benefits and costs of the ETS in the EU, a lesson learned for the CBAM design*. ECB Working Paper Series No 2764. https://www.ecb.europa.eu/pub/pdf/scpwps/ecb.wp2764~3ff8cb597b.en.pdf

Burke, J., Sato, M., Taylor, C., & Li, F. (2021). *What does an EU carbon border adjustment mechanism mean for the UK?* Grantham Research Institute on Climate Change and the Environment and Centre for Climate Change Eco-nomics and Policy, London School of Economics and Political Science. https://www.lse.ac.uk/granthaminstitute/wp-content/uploads/2021/04/What-does-an-EU-Carbon-Border-Adjustment-Mechanism-mean-for-the-UK_FULL-REPORT.pdf

CENEF-XXI. (2021). CBAM: Implications for the Russian economy (In Russ.). https://cenef-xxi.ru/uploads/Cz_ENEF_XXI_CBAM_4c0a2fb4a3.pdf

Chepeliev, M. (2021). Possible implications of the European carbon border adjustment mechanism for Ukraine and other EU trading partners. *Energy Research Letters, 2*(1). https://doi.org/10.46557/001c.21527

Council of the European Union, Regulation of the European Parliament and of the Council establishing a carbon border adjustment mechanism (CBAM)—Compromise text, 16060/22, 14 December 2022. https://data.consilium.europa.eu/doc/document/ST-16060-2022-INIT/en/pdf

European Commission. (2021). Proposal for a regulation of the European Parliament and of the council establishing a carbon border adjustment mechanism. https://eur-lex.europa.eu/legal-content/EN/TXT/?uri=celex:52021PC0564

IEA, Imperial College Business School. (2021). Clean energy investing: Global comparison of investment returns. https://www.iea.org/reports/clean-energy-investing-global-comparison-of-investment-returns

IPEM. (2021). Analytical report "Cross-border carbon regulation in the EU: How to prevent discrimination against Russian exporters". https://ipem.ru/content/transgranichnoe-uglerodnoe-regulirovanie-v-es-kak-ne-dopustit-diskriminatsii-rossiyskikh-eksportyerov/

Kaitila, V., Kuusela, O.-P., Kuusi, T., Pohjola, J., & Soimakallio, S. (2022). *Revisiting the economic impacts of the EU CBAM on Finland and the EU*. ETLA Reports 128, The Research Institute of the Finnish Economy. https://www.etla.fi/wp-content/uploads/ETLA-Raportit-Reports-128.pdf

Kuehner, A.-K., Jakob, M., & Flachsland, C. (2022). German stakeholder perceptions of an EU carbon border adjustment mechanism. *Environmental Research Letters, 17*, 124007. https://doi.org/10.1088/1748-9326/ac9f23

Lim, B., Hong, K., Yoon, J., Chang, J.-I., & Cheong, I. (2021). Pitfalls of the EU's carbon border adjustment mechanism. *Energies, 14*, 7303. https://doi.org/10.3390/en14217303

Magacho, G., Espagne, É., & Godin, A. (2022). *Impacts of CBAM on EU trade partners: Consequences for developing countries*. AFD Research Papers. https://www.cairn-int.info/journal%2D%2D2022-238-page-1.htm.

Marcu, A. (2021). EU CBAM proposal (Presentation). https://cenef-xxi.ru/uploads/Session_1_A_Marcu_EU_CBAM_Proposal_a5c6b40d6e.pptx.

Marcu, A., Maratou, A., Mehling, M., & Cosbey, A. (2021). The EU carbon border adjustment mechanism (CBAM). Preliminary analysis of the European Commission proposal for a regulation establishing a carbon border adjustment mechanism. https://ercst.org/the-eu-cbam-preliminary-analysis-fit-for-55/.

Qi, S., Xu, Z., & Yang, Z. (2022). China's carbon allowance allocation strategy un-der the EU carbon border adjustment mechanism: An integrated non-parametric cost frontier approach. *Sci Total Environ, 831*, 154908. https://doi.org/10.1016/j.scitotenv.2022.154908

Schott, J. J., & Hogan, M. (2022). *Is South Korea vulnerable to EU and US carbon border restrictions?* Peterson Institute for International Economics Policy Brief. https://www.piie.com/sites/default/files/documents/pb22-10.pdf

Siy, A. L., Wang, A., Zheng, T., & Hu, X. (2023). Research on the impact of the EU's carbon border adjustment mechanism: Based on the GTAP model. *Sustainability, 2023*(15), 4761. https://doi.org/10.3390/su15064761

Takeda, S., & Arimura, T. H. (2023). *A computable general equilibrium analysis of EU CBAM for the Japanese economy*. RIETI Discussion Paper Series 23-E-006. https://www.rieti.go.jp/jp/publications/dp/23e006.pdf

Zhong, J., & Pei, J. (2022). Beggar thy neighbor? On the competitiveness and welfare impacts of the EU's proposed carbon border adjustment mechanism. *Energy Policy, 162*, 112802. https://doi.org/10.1016/j.enpol.2022.112802

Zimmer, M., & Holzhausen, A. (2020). *EU carbon border adjustments and developing country exports: Saving the worst for the last*. Allianz. https://www.allianz.com/content/dam/onemarketing/azcom/Allianz_com/economic-research/publications/specials/en/2020/november/2020_11_17_EU_CBAM_and_develping_country_exports.pdf

Chapter 5
Inclusive Growth of National Economies and New Energy Paradigm

Bulat Khusainov, Asset Nussupov, Sholpan Kaimoldina, and Baurzhan Temirbayev

Abstract The aim of the study is to assess and analyze the impact of energy consumption on the inclusive growth of European and Asia-Pacific economies in the context of the new paradigm of energy transition. The novelty of the study consists, firstly, in quantifying the inclusive growth of 55 economies over a 20-year retrospective period. Secondly, it is one of the first studies to assess the impact of consumption of different types of energy resources, reflecting the challenges of the energy transition, on the inclusive growth of the economies of Europe, Central Asia, and Asia-Pacific. The methodology for constructing the inclusive growth index is based on the principal component method, and the assessment of causal relationships between it and energy consumption indicators is based on three econometric models using panel data. Results (a) the main hypothesis of the study on the impact of energy consumption on the inclusive growth of national economies is quantitatively confirmed; (b) the multidirectional impact of six types of energy resources on the inclusive growth of the economies of Europe, Central Asia and Asia-Pacific countries is revealed. Conclusions on the prospects of implementation of the new energy paradigm in the countries of these regions are presented.

5.1 Introduction

The second decade of the twenty-first century witnessed global challenges like Brexit, trade wars between the USA and China, COVID-19 coronavirus pandemic. Earlier, there were plenty of constant statements of "the end" of globalization and dawn of deglobalization era in the modern world development agenda for politicians and economists. However, a careful analysis of the indicators that traditionally

B. Khusainov · A. Nussupov (✉) · S. Kaimoldina · B. Temirbayev
Zhetysu University, Taldykorgan, Kazakhstan
e-mail: b.khusainov@zu.edu.kz

© The Author(s), under exclusive license to Springer Nature Switzerland AG 2024

T. Devezas et al. (eds.), *Global Energy Transition and Sustainable Development Challenges, Vol. 1*, World-Systems Evolution and Global Futures,
https://doi.org/10.1007/978-3-031-67583-6_5

measure globalization does not convince us in globalization dusk. Rather, we see an inevitable slowdown after a period of hyperglobalization—so called "slowbalization" ("slow" and "globalization"). Moreover, deep changes of the political environment and public sentiment in the largest economies over the past 5 years could be a good start for a new era, which will possibly lead to a slowdown in economies, rising inequality and, perhaps, new conflicts (Goldberg & Reed, 2023).

Modern global development demonstrates the impact of many factors on national economies and their policies. In recent decades, along with economic factors, its social and environmental components have become increasingly important. There is a special place among them taken by inclusiveness of growth and the new energy paradigm. At first glance, it has simple essence—the transition to new, "clean" energy sources. However, introduction to the reality of the new energy paradigm is multitask and extremely complex problem that requires significant multidimensional costs (economic, financial, technological, etc.) and all world states' efforts. It is clear that the transformation based on the paradigm of energy transition requires a large-scale expansion of the use of renewable energy sources (RES). At the same time, this determines the need to increase energy efficiency to meet the rapidly growing demand of end-users of energy. The new energy paradigm, of course, should take into account the needs of citizens and communities after abandoning the use of fossil fuels, and provide millions of people with access to modern energy sources (World Bank, 2023). It should also be aimed at a fair solution to a number of other sensitive issues.

Some researchers believe that the new energy paradigm will become a driver of economic growth. Correctness of the statement will be clear after the implementation of the energy transition. However, economic growth is not a goal in itself. In fact, it is one of the necessary conditions for population's welfare improvement. But growth quality is more important than its pace (Khusainov et al., 2022). It seems that inclusiveness is the key indicator of growth quality, that is, the even distribution of benefits from economic growth among various groups of the population and the improvement of their well-being.

Effective use of the economic potential is essential for increasing demand of the population that leads to rising debates on these issues. The cascade of available information on the new energy paradigm topic, inclusive growth issues, as well as their cause-and-effect relationships can create cognitive distortions after inadequate interpretation of such complex economic phenomena and processes. In this context we consider the energy transition that meets the rapidly growing demand for energy alongside should be accompanied by inclusive growth of national economies. The obvious unevenness and ambiguity for achieving carbon neutrality by the world countries will certainly result in increase of not just a global inequality, but also income heterogeneity at the intercountry and national levels. The tectonic faults in the global economy and the geopolitical turbulence that occurred after February 2022, associated with well-known events, significantly increase the need for serious study in this field. We made the substantial analysis of many related publications and found few that investigate the causality between inclusive growth and consumption of various energy sources by regions and world countries. This determines the relevance and novelty of the given study.

5.2 The New Energy Paradigm and Inclusive Growth

The promotion of fair and equitable economic growth was the key macroeconomic issue discussed in recent decades. All countries of the world challenge finding a compromise between social security and the strong growth of their economies. The study of the dynamics of the relationship between social security and economic growth stimulates a better understanding of the causes of income inequality. Identifying these relationships facilitates improvement in defining policies that promote inclusive growth and reduce concerns about income stratification. The refrain of many studies is the idea that the desire to achieve high economic growth through a series of political reforms catalyzes the increase in inequality and neutralize (destroy) inclusive growth.

There are several serious incentives for shifting the focus of research to the inclusiveness of the growth of national economies, especially in the context of the new energy paradigm. First, the gross domestic product (GDP), despite its importance, is not an adequate indicator of the income and the country's population welfare (Khusainov et al., 2022; Stiglitz et al., 2009). Secondly, traditional economic growth can't stop the stratification of the population in terms of income and wealth, does not guarantee universal and equal access to economic benefits. Thirdly, ambiguity of the perspectives of energy transition announced by many countries is caused by the coronavirus pandemic of 2020–2022, the trade war between the United States and China, which began in 2018 with continuing echoes, generate turbulence in global development. Having this background, the situation sharply worsened in February 2022 and led to a tectonic fracture of the modern world order, giving rise to greater uncertainty on a planetary scale in socioeconomic and political development. The noted economic shocks and conflicts worsen the macroeconomic conditions of development, giving rise to "instability traps" (Patcharaporn et al., 2023). One of the world's leading economists on inequality, B. Milanovich, is convinced that the era of reducing global inequality due to the convergence of countries in terms of income, which lasted since the early 1980s of the last century, is coming to an end. The "China's effect", whose rapid economic growth has served as the main "leveling force" since the late 1970s, is drying up, while the growth of other countries is unlikely to be able to make a comparable contribution. In his opinion, we are facing the most serious defeat on the path of reducing poverty and inequality since the Second World War (Milanovich, 2022, pp. 25, 32). For these reasons, we consider there is no doubt that the countries of the world need a policy aimed at inclusive growth of their economies, since it increases population incomes and decreases inequality. We should add that the energy transition and related problems should be considered in the context of the decrease in inequality (Grigoryev et al., 2020), and therefore in the context of inclusive growth.

A New Energy Paradigm Since the 1970s, the global energy policy has been based on the "carbon paradigm". At present, the transition from carbon-based to climate-neutral energy is an objective necessity (Borovsky, 2021). The realities of modern development predetermine the extreme urgency and necessity of transition to a new energy paradigm. Many studies have been devoted to this issue, a comprehensive review of which is presented, for instance, in Parra-Domínguez et al. (2023). Without emphasizing them, let us briefly review approaches to the implementation of the new energy transition in the regions and countries of the world that are the focus of this study. Thus, the EU has developed a Green Deal development strategy, which aims to build a carbon-neutral economy by 2025 (Vlahinić Lenz & Fajdetić, 2021). After conducting a study on 26 EU countries for the period between 2000 and 2018, the authors came to the following conclusion. One of the negative effects of globalization is the increasing emissions of greenhouse gases. According to them, environmental taxes, as an important policy instrument, can reduce the negative impact of greenhouse gas emissions on the environment and climate change. However, the opposite impact of environmental taxes has been identified for a number of EU countries. The results of the study, obtained by the authors, confirm the correctness of the Green Deal strategy chosen by the EU. It is clear that the findings of this study are not applicable to other countries and/or regions due to differences in economic, political, social, and other development factors.

Eurasian states also have national low-carbon development strategies. In Russia, for example, it is the Strategy of Socio-Economic Development with Low Greenhouse Gas Emissions until 2050. In the paper of (Porfiriev & Shirov, 2022) authoritative Russian scientists, formulated principles and approaches to reduce the risks of decarbonization of the economy by emphasizing the key role of the global climate agenda. In 2023 Kazakhstan adopted the Strategy for Achieving Carbon Neutrality by 2060 (2023).

It is well known that China is the main polluter of the environment and is restructuring its energy system to achieve the "30–60" decarbonization goal (Dong et al., 2021). It is predicted that the country will reach peak CO_2 emissions by 2023 and then decline, reaching carbon neutrality by 2060 (Huo et al., 2021). In recent years, China has accelerated the process of replacing fossil fuels. It should be added that in early 2021, the authorities of the Celestial Empire called for the construction of a new power system (NPS) using new energy as the main source, which is characterized by a high proportion of renewable energy sources.

East Asian countries have also been developing and updating national carbon neutrality strategies for decades. Thus, a key element of Japan's New Growth Strategy (NGS) is to strengthen the country's industrial and technological capabilities in seven strategic areas by 2030. First on the list are environment and energy ("green innovation"). The main objective of the NGS is to create a low-carbon society through an integrated policy package (systemic change, new regulations, and regulatory reform) and to support the rapid diffusion of environmental technologies and products. The Republic of Korea aims to turn the country into a "fifth green power" by 2050. In particular, the "Land of the Morning Calm" under the East Asia

Climate Partnership initiative links national development goals (improving quality of life) with broader international low-carbon development efforts (Dent, 2012).

For the countries of South-East Asia (Indonesia, Malaysia) the idea of transition to the trajectory of sustainable development and "green" growth became a springboard (base) for strategic planning documents (the principles of sustainable development are integrated into 5-year plans). The energy transition is being actively promoted for key economic sectors selected in line with national priorities. For the aforementioned countries, the policy is aimed at long-term changes in the structure of the national economy in favor of energy transition, energy saving, and growth in the share of high-tech goods and services. In general, countries use the global sustainable development agenda as an opportunity to modernize the national economy and implement reforms, as well as an argument to justify the need for transformation to national elites (Yakovlev et al., 2020).

To conclude this brief overview of the new paradigm of the energy transition, we note that the number of countries that are actively promoting the energy agenda is constantly growing. If by 2005 the number of countries that had set the bar for diversifying growth with renewable energy did not exceed 55, by 2011 the number had already risen to 100 (Dent, 2012). In 2021, the number of countries aiming to achieve carbon neutrality amounted to 130 (71% of global greenhouse gas emissions), then as of March 2023, this number had increased to 140 (90% of greenhouse gas emissions).

Inclusive Growth: Theoretical Views and Measurement Practice In the scientific literature, there is a myriad of publications with abundant views on the phenomenon of "inclusive growth", an overview of which, as a rule, can be found in most of them. Inclusive growth studies are conducted both at the level of specific countries (Che Sulaiman et al., 2021; Rytova et al., 2021; Sharafutdinov et al., 2019) and by groups of countries (Heshmati et al., 2019) and continents (Popкova et al., 2020, pp. 123–182). In the context of our research, we will focus on works aimed at study of the inclusiveness of growth at the intercountry level.

The issue of sustainable and inclusive growth is reflected in the G-20 agenda, which covers a wide range of issues, including macroeconomics, finance, healthcare, energy transition, digital economy, education and science, etc. It is emphasized that in the long-term ensuring inclusive growth and citizens welfare as the priorities of the G-20 should help vulnerable countries, contribute to the conservation of nature and climate and ensure equal opportunities for all in the era of the digital economy (Lukash, 2021, pp. 7–14).

The work (Stawska & Jabłon'ska, 2022) identifies the level of inclusive growth in the EU-27 in 2000, 2008, and 2020, which measures are determined by the corresponding index based on the method of weight correlation, and assessment of its determinants impact on the economic and financial development of the European Union member countries. Along with economic and social indicators of development, the authors placed special emphasis on factors that influence inclusive growth by the government and central banks. As a result of the study, it was found that inclusive growth in each EU-27 country is influenced by many factors that cannot be combined in one indicator. These factors are also influenced by the decisions of governments, central banks and other governing bodies. Perhaps the key conclusion of the authors can be formulated as follows: in order to ensure inclusive growth in most EU countries, fundamental reforms must be carried out. Collective contribution to socioeconomic development at all levels is also needed.

The analysis of Central and Eastern Europe (CEE) policy based on inclusive growth and development models is presented in the work (Tsapko-Piddubna, 2021). It has been revealed that redistributive fiscal policy has little effect on inclusive growth and development of CEE national economies. The main conclusion of the study is that individual countries in the region have not transformed well into social integration. It is emphasized that the CEE countries need to focus on inclusive growth policies. This, according to the authors, will increase equality and economic well-being, reduce poverty, vulnerability and marginalization of certain segments of the population. Ultimately, this will increase the competitiveness of the region in the global economy.

Many researchers consider economic growth as a mechanism that at the initial stages can contribute to increasing income inequality, but in the long term it still reduces the level of poverty (Atkinson, 2018; Khusainov et al., 2021; Piketty, 2018; Stiglitz, 2015, 2020). In order to be sustainable and effective in reducing poverty and inequality growth must be inclusive (OECD, 2018a). For countries with any level of economic development, the ability to stimulate growth on an equitable basis is the main priority of national economic policy. It seems fair to say that inclusive growth, whatever its interpretation, refers to the distribution of growth and fairness, and not necessarily to the speed and level of growth.

In this context, we consider one of the indirect criteria for assessing inclusive growth could be inequality in income and wealth. For example, in 2021, the poorest half of the population earned 8.2% of world income (in 1995—7.0%), and the richest 10% of the world's population received 52.5% (56.2%) of it. Global wealth inequality is even more prominent: in 2021, the poorest half of the world's population owns only 1.7% (1.4%) of all accumulated wealth. On the contrary, the richest 10% of the world's population accounted for 77.1% (79.9%) of all wealth (Fig. 5.1). The differentiation in income and wealth among these groups of the adult population is even more significant by regions and states (Chancel et al. 2022).

The problem of inclusive growth is not only in the sphere of scientific interests of researchers from various countries, but also of many international organizations. Thus, the World Bank (WB), within its own concept, defines economic growth as inclusive if its inclusiveness is observed along with the growth of real GDP, and lead to an increase in productive employment (i.e., an increase in both employment and labor productivity growth), poverty reduction and allows socially excluded people to enjoy the benefits of economic growth and much more (Commission on Growth and Development 2008). The Asian Development Bank (ADB) interprets inclusive growth in a broad sense (inclusive growth as economic growth that not only creates new opportunities in the economy, but also provides equal access to them for all groups of the population, especially for the poorest) and in a narrow sense—If, firstly, the incomes of all groups of the population increase due to their

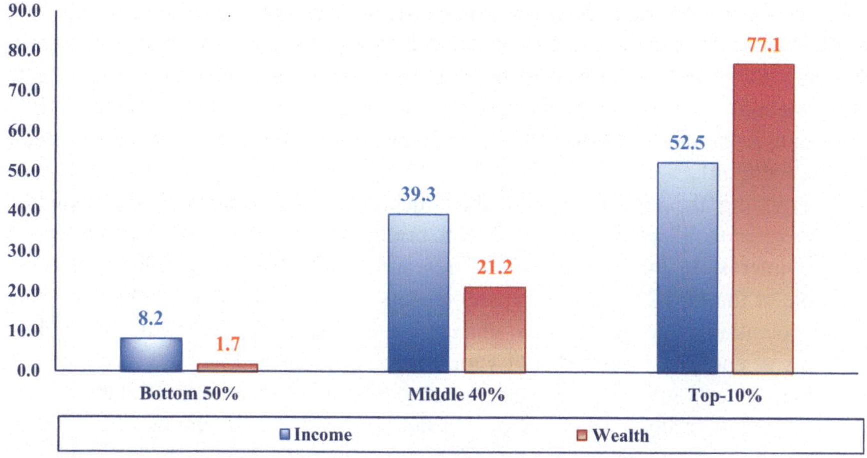

Fig. 5.1 Global income and wealth inequality, 2021. Source: compiled by the authors according to wid.world/data. Interpretation: The global bottom 50% captures 8.2% of total income measured at Purchasing Power Parity (PPP). The global bottom 50% owns 1.7% of wealth (at PPP). The global top 10% owns 77.1% of total Household wealth and captures 52.5% of total income in 2021. Note that top wealth holders are not necessarily top income holders. Incomes are measured after the operation of pension and unemployment systems and before taxes and transfers

"participation"; secondly, "non-monetary" (related to education, healthcare, social integration, etc.) disparities between different groups of the population are decreasing. In other words, ADB focuses equally on both results and the growth process (Aizenman et al., 2012).

One of the three priorities of the EU Development Strategy ("Europe 2020") is to achieve inclusive development. According to the concept of the European Commission, inclusive growth is achieved by ensuring a high level of employment, investing in the acquisition of necessary skills, combating poverty and modernizing the labor market, in order to form a cohesive society. Moreover, a prerequisite for achieving inclusive growth is to spread the positive effects of growth to all regions of the EU. It is assumed that such an approach will help to equalize the quality of life even in remote and less developed regions of the European Union (European Union, 2010).

The Organization for Economic Cooperation and Development (OECD) defines inclusive growth as growth that is fairly distributed throughout society and creates opportunities for all, as well as its potential to reduce increasing income heterogeneity (OECD, 2018b). Various concepts of inclusive growth and methods of its measurement are also offered by other international organizations: United Nations Conference on Trade and Development (UNCTAD, 2018), World Economic Forum (WEF, 2018), etc.

As a rule, inclusive growth is studied in relation to various aspects of the socioeconomic development of states. At this background we found an interesting study (Smorgunov, 2018), that considers the inclusiveness of growth with administrative reform in the BRICS countries (Brazil, Russia, India, China, and South Africa). As the henchmen of the concept of an inclusive growth strategy based on the unity of manageability, competitiveness and public policy through civic participation, the authors are convinced that this approach is fair enough. Indeed, strategic public policy, as well as the system of public administration, will not be effective and inclusive if there is no adequate institutional environment (while reforming its economy China, first of all created the institutions necessary for their implementation).

Concluding the analysis of the theoretical and methodological views on the "inclusive growth" phenomena, we will briefly focus on two works that are of special interest in the context of our research (Grigoryev et al., 2020) use the example of the BRICS countries and considers the principles and mechanisms that could potentially interlink climate change issues, including caused by the carbon intensity of energy consumption of various social groups and income inequality. In fact, the authors advocate the need to introduce a differentiated climate tax for various social groups, which income would be distributed among the poor to mitigate the problem of inequality. They are convinced that such a system will allow countries to reduce the damage to the poor or middle-income segments of the population in developing countries, as well as not exacerbate the problem of poverty or restrain economic growth itself.

Even more interesting in terms of our research is perhaps a single work (Cui et al., 2022), which shows with a mathematical toolkit the great impact of the expansion of financial inclusion and the consumption of renewable energy sources (RES) on promotion of inclusive growth.[1] In particular, the authors study the impact of financial services coverage and renewable energy consumption on global inclusive growth at the example of 40 world countries. They also analyzed the impact of modernization of the industrial structure, technological progress, government loans, population density, and education levels on the inclusiveness of growth. An important conclusion is that the modernization of the industrial structure and technological progress contribute to the inclusive growth of countries, while government loans, on the contrary, have a negative impact.

The most common way to measure the inclusiveness of growth is to build a composite index, which is differently constructed in international organizations and in academic research. Set of used indicators and the mathematical apparatus for their construction and calculation formulate the difference between inclusive growth indices. It is important to make a small remark here. In recent decades, the governments of many states and, as a result, their researchers, have been experiencing religious reverence for various indices. Their use conceals both positive and negative aspects. The fundamental mistake is the false idea that composite indexes, as opposed to use individual indicators, make it easier and better to understand trends in the economy, social sphere, climate change, etc. However, incorrectly constructed composite indexes and/or their incorrect interpretation can lead to simplified political conclusions, their use to support the desired policy, and finally mislead. All this

[1] The idea of our research appeared much earlier than we found (Cui et al., 2022) publication while searching SCOPUS and WoS international databases.

threats the adoption of inadequate erroneous decisions in economic policy. Ideally, a composite index should measure multidimensional concepts that cannot be covered by a single indicator, such as competitiveness, industrialization, sustainability, inclusivity, etc. In other words, they allow us to effectively generalize multidimensional realities of economic processes and phenomena, use them for a comparative analysis of the development of various countries. Given the complexity of the phenomenon of inclusive growth, it seems reasonable to use a consistent composite index in this study. It should be added that various methods of multidimensional analysis are used for composite index construction, such as: principal component analysis, factor analysis, Cronbach's Alpha coefficient, cluster analysis, etc. (OECD, 2008).

5.3 Methodology and Information Base

The study was conducted in two stages using economic and mathematical toolkit. During the first stage, Principal Components Analysis (PCA) was used to calculate composite indexes of inclusive growth of 55 national economies; during the second stage, econometric models were used to identify the causality between the inclusive growth of economies and the consumption of various types of energy resources.

The PCA method's main aim lies in explaining the variance of the observed data by using several linear combinations of the raw data (OECD, 2008, pp. 63–69, pp. 89–91). The composite Inclusive Growth Index (IGI) was calculated using the *Python* programming language based on a set of indicators grouped into four pillars for which sub-indexes were calculated step by step.

The first step is to conduct a correlation analysis between the variables because the PCA method necessitates the existence of a correlation between the variables. Otherwise, its use will be impractical. In the second step, principal components were selected using Kaiser's rule and graphical method. Standard practice suggests selecting factors with eigenvalues greater than 1 that together explain more than 60% of the total variance. After determining the number of factors, their rotation was carried out. Following the accepted practice, a varimax rotation was applied, which aims to minimize the number of variables with high loadings on each factor. This method simplifies the interpretation of the factors. The feasibility of factor analysis was tested by Bartlett's sphericity criterion and the Kaiser-Meyer-Olkin criterion.

The following step (Dharmaratne & Attygalle, 2018) calculated the weights of the variable based on the obtained factor matrix after rotation. They are the normalized squares of the factor loadings and are calculated as the proportion of the square of the factor loading to the total unit variance of the indicator:

$$L_i = \frac{(F_i)^2}{V_j}, \quad (5.1)$$

where L_i—normalized square of factor loading of variable i; F_i—factor loading of variable i; V_j—variance explained by factor j.

As a result, the variables with the highest weights are grouped into intermediate indicators that have different degrees of importance. Their weights were then calculated as the proportion of variance explained in the dataset:

$$F_j = \frac{V_j}{\sum V_n}, \quad (5.2)$$

where F_j—weight of factor j.

The final weights for each variable are then calculated using the following formula:

$$W_i = L_i * F_j, \quad (5.3)$$

where W_i—weight of variable i.

In the final step, the IGI subindex is calculated, which is the weighted sum of the normalized variables:

$$I = \sum_{i=1}^{n} X_i W_i, \quad (5.4)$$

where I—IGI subindex; X_i—normalized variable i, calculated by the formula $\frac{X_i - X_{min}}{X_{max} - X_{min}}$.

The second step assessed and identified the relationships between inclusive growth and energy consumption in the analyzed countries, representing different regions and worlds dominated by different energy sources. In particular, the econometric toolkit used estimates inclusive growth as a linear function of energy consumption indicators for different groups of economies in two large regions. The equations are constructed using panel data: the original regression specification takes the following form:

$$\text{Log } Y_{i,t} = a + b_1 \text{ Log}X_{i,1} + b_2 \text{ Log}X_{i,2} + \ldots + b_n \text{ Log}X_{i,n} + \varepsilon_{i,t} \quad (5.5)$$

where i—country, t—time period. The variable is Log$Y$ the logarithm of the real value of the dependent variable, $X_{i,n}$—consumption of oil, gas, coal, RES, hydro and nuclear energy, as well as specific CO_2 emissions.

As part of the analysis, the generalized least squares method (LSM) was used. All results are based on the estimation of an "LSM framework" that exploits both between-country and within-country variation in variables across the three models. First, a panel data model with individual fixed effects. Relies on a panel data structure to account for unmeasured individual differences in objects. These differences are known as effects. In this model, effects are interpreted as an interfering parameter; the estimation aims to exclude them. Second, a panel data model with time fixed effects. It eliminates systematic error due to unobserved quantities that change over time but remain constant for objects. The model controls for factors that vary between sites but remain constant over time. It is clear that in fixed-effect and non-fixed-effect models, the results can differ substantially. Third, a panel data model with two-way fixed effects, i.e., synthetic (integral) use of individual fixed effects and time fixed effects. Thus, the application of the three different models accounts for the main source of variation in energy consumption variables (i.e., cross-country differences).

Data Quality The study focuses on 55 countries in two regions of the world. The first group is the Europe and Central Asia (CA) region, which includes 58 countries according to the WB classification, but 43 countries are included in our study (availability of qualitative dynamic series). The second group includes 12 countries in East (8 countries), South Asia (2 countries), and the Pacific (2 countries). The 17 indicators (Table 5.1) available primarily in the World Development Indicators (WDI) and World Inequality Data (WID) were used to construct an inclusive growth index. The choice of countries and variables is due to the availability of qualitative dynamic series covering a 20-year period (2000–2019).[2] All indicators were grouped into four pillars: pillar 1—economy, pillar 2—standard of living, pillar 3—employment and inequality, and pillar 4—Worldwide Governance Indicators (WGI).

In the second part, the focus of the study is on the consumption (rather than production) of energy resources and, as a consequence, CO_2 emissions. Statistics on the consumption of all energy sources (oil, gas, coal, hydro, nuclear, RES) in exajoules[3] and absolute CO_2 emissions in million tons ($MtCO_2$) are taken from the bp statistical Review of World Energy 2022 database (bp Review, 2022), and specific CO_2 emission rate (kg per 2015 USD of GDP) from WDI.[4] The number of countries for the two study regions has changed because in (bp, 2022) the data for some countries are aggregated into "other" country groups for the respective regions and continents. Thus, in the Europe and Central Asia region, energy consumption data for 34 countries were used, and in the East, South Asia and Pacific group of countries—for 12 countries.

[2]The dynamic series ends in 2019, as for one of the important pillar 3 indicators, data for 2020–2021 are not published at the time of the study.

[3]1 exajoule (EJ) = 1 quintillion joules (1×10^{18}).

[4]In WDI, carbon dioxide emissions are emissions resulting from the burning of fossil fuels and cement production. They include carbon dioxide produced by consuming solid, liquid, and gaseous fuels and flaring gas.

Table 5.1 Indicators for calculating the composite index of inclusive growth

Pillar	Indicators		
Pillar 1	1.1	GDP per capita PPP (constant USD 2017)	
	1.2	Net national income (NNI) per capita PPP (constant USD 2021)	
	1.3	Labor productivity, USD/person (GDP per person employed (constant 2017 PPP USD))	
	1.4	Employment rate (ratio to labor force), 15+, total (%) (modeled ILO estimate)	
	1.5	Exports of goods and services (% of GDP)	
Pillar 2	2.1	Final consumption expenditure (% of GDP)	
	2.2	Government spending on health (% of GDP)	
	2.3	Life expectancy at birth, total (years)	
	2.4	Individuals using the Internet (% of population)	
Pillar 3	3.1	Self-employed, total (% of total employment) (modeled ILO estimate)	
	3.2	Net Personal Wealth, Bottom 50%, share adults	
Pillar 4	Worldwide Governance Indicators (WGI)		Description
	4.1	Voice and accountability	Perception of the extent to which a country's nationals can participate in government election, as well as freedom of expression, association, and free media.
	4.2	Political stability and absence of violence/ terrorism	An index that maps perceptions of the likelihood that a government will be destabilized or overturned by unconstitutional measures or violence, including politically motivated violence and terrorism.
	4.3	Government effectiveness	Perception of the quality of public services, the quality of the civil service and its degree of independence from political pressures, the quality of policy formulation and implementation, and the credibility of government involvement in such policies.
	4.4	Regulatory quality	Perception of a government's ability to formulate and implement sound policies and regulations that enable and promote private sector development.
	4.5	Rule of law	Perceptions of the extent to which agents trust and adhere to the rules of society, in particular the quality of contract enforcement, property rights, the police and the courts as well as the likelihood of crime and violence.
	4.6	Control of corruption	Perception of the extent to which public power is exercised for private gains, including both minor and major forms of corruption, as well as elite "seizure" of the state and private interests.

Source: compiled by the authors

5.4 Inclusivity and Energy Resources: An Econometric Analysis

Let us preface further exposition with a remark on some of the differences of the composite IGI used in our study. Firstly, traditionally, international organizations and academia calculate IGI in a point-by-point manner, e.g., related to the crisis year (s). In other words, "point" inclusiveness is defined. Secondly, this is partly why the set of indicators used to calculate IGI varies from dozens to more than a hundred.

The principal novelty of the present development is expressed in the fact that the IGIs are constructed for a relatively long, 20-year period, which is predetermined by the purpose and objectives of the study. Therefore, the main criterion for selecting variables to determine IGI was the availability of completeness of qualitative data for the dynamic series used. It is clear that the calculated IGIs do not include a number of extremely important economic and social indicators, such as: expenditures on education, expenditures on research and development, etc. However, the statistical series for the aforementioned and other relevant indicators not included in the calculations are discrete. The lack of environmental indicators may be reproached, but one of them (CO_2 emissions) is deliberately not included in the IGI calculations, because it is involved in the econometric analysis to identify the causality of the inclusive growth index and energy consumption.

Another novelty is the use of a non-traditional indicator Net Personal Wealth, Bottom 50%, share adults (pillar 3.2). It should be added that another non-traditional indicator—Net Personal Income, Bottom 50%, share adults—was excluded due to collinearity with the previous share indicator of the wealth of the poorest half of the population, which makes the IGI difficult to estimate and analyst.

An analysis of IGIs calculated for Europe and CA shows that the Netherlands had the highest value in 2019; Luxembourg was in second place, followed by Iceland in third place. Developed European countries outperformed the region's underdeveloped economies, including the post-Soviet ones, being in the top half of the list ranked by IGI value throughout the period (Table 5.2).

In 2019, the trailing ten countries, starting from the outsider position, included: Uzbekistan, Russia, Ukraine, Kyrgyzstan, Kazakhstan, Turkey, Bosnia and Herzegovina, Bulgaria, Moldova, and Armenia. It is clear that the level of socioeconomic development, quality of life, and a number of other key indicators of advanced EU countries are much higher, as well as the distribution of benefits from growth, compared to CEE economies, as well as European and Central Asian post-Soviet states.

A similar trend of growth inclusiveness is observed in East Asia, South Asia, and the Pacific (Table 5.2). Throughout the whole period, the region's four advanced economies—Australia, Japan, New Zealand, and the Republic of Korea—represent the leading four. In various years, Indonesia, Thailand, Pakistan, Philippines, and Thailand are in the trailing quartet in terms of inclusiveness. China is consistently in the middle ranks.

Table 5.2 Indexes of inclusive growth for Europe and Central Asia, East, South Asia, and the Pacific countries, 2000–2019

Country/year	2000	2001	2002	2003	2004	2005	2006	2007	2008	2009	2010	2011	2012	2013	2014	2015	2016	2017	2018	2019
Europe and Central Asia																				
Netherlands	0.60	0.60	0.61	0.60	0.60	0.61	0.60	0.60	0.61	0.61	0.61	0.62	0.61	0.60	0.63	0.66	0.67	0.67	0.66	0.70
Luxembourg	0.63	0.64	0.64	0.63	0.65	0.65	0.65	0.65	0.65	0.66	0.66	0.66	0.64	0.64	0.66	0.67	0.67	0.67	0.67	0.67
Iceland	0.63	0.62	0.64	0.64	0.64	0.65	0.63	0.62	0.62	0.60	0.59	0.60	0.60	0.60	0.62	0.63	0.63	0.64	0.63	0.65
Sweden	0.57	0.57	0.58	0.58	0.59	0.59	0.58	0.58	0.59	0.59	0.58	0.61	0.61	0.61	0.63	0.64	0.64	0.64	0.62	0.64
Denmark	0.57	0.56	0.56	0.58	0.59	0.61	0.61	0.61	0.61	0.59	0.59	0.59	0.59	0.60	0.61	0.63	0.62	0.62	0.60	0.63
Switzerland	0.60	0.59	0.59	0.58	0.58	0.59	0.58	0.58	0.59	0.58	0.58	0.59	0.59	0.59	0.61	0.61	0.62	0.62	0.60	0.61
Belgium	0.56	0.56	0.57	0.56	0.56	0.57	0.57	0.57	0.57	0.57	0.58	0.59	0.60	0.61	0.62	0.63	0.63	0.62	0.60	0.61
United Kingdom	0.54	0.54	0.55	0.55	0.57	0.57	0.57	0.57	0.58	0.56	0.57	0.57	0.57	0.57	0.60	0.62	0.62	0.61	0.59	0.61
Norway	0.54	0.54	0.54	0.53	0.54	0.55	0.54	0.54	0.55	0.55	0.55	0.55	0.56	0.57	0.58	0.60	0.61	0.61	0.59	0.60
Germany	0.55	0.54	0.54	0.53	0.53	0.55	0.55	0.54	0.54	0.54	0.54	0.55	0.55	0.56	0.59	0.60	0.60	0.59	0.57	0.59
Finland	0.55	0.55	0.55	0.55	0.55	0.56	0.56	0.55	0.57	0.57	0.57	0.58	0.58	0.59	0.61	0.61	0.60	0.59	0.57	0.59
Austria	0.53	0.53	0.53	0.52	0.53	0.54	0.54	0.55	0.56	0.54	0.54	0.55	0.55	0.56	0.58	0.58	0.58	0.59	0.57	0.59
France	0.52	0.52	0.52	0.51	0.53	0.54	0.54	0.54	0.55	0.53	0.54	0.54	0.54	0.55	0.56	0.57	0.57	0.57	0.55	0.56
Spain	0.52	0.52	0.51	0.52	0.52	0.54	0.52	0.51	0.52	0.50	0.51	0.52	0.51	0.51	0.52	0.54	0.55	0.55	0.52	0.54
Slovenia	0.47	0.48	0.48	0.47	0.49	0.50	0.51	0.50	0.51	0.51	0.52	0.52	0.51	0.51	0.53	0.53	0.53	0.53	0.51	0.53
Czechia	0.40	0.41	0.44	0.44	0.44	0.45	0.46	0.45	0.47	0.47	0.47	0.48	0.48	0.49	0.51	0.52	0.52	0.52	0.50	0.51
Ireland	0.48	0.48	0.48	0.47	0.48	0.50	0.51	0.51	0.53	0.52	0.52	0.51	0.50	0.50	0.51	0.50	0.49	0.49	0.49	0.50
Slovak Republic	0.43	0.43	0.45	0.45	0.48	0.49	0.50	0.50	0.51	0.51	0.51	0.51	0.50	0.50	0.50	0.51	0.52	0.50	0.48	0.49
Cyprus	0.50	0.49	0.50	0.50	0.49	0.50	0.51	0.50	0.52	0.51	0.52	0.52	0.50	0.50	0.51	0.50	0.49	0.49	0.47	0.49
Portugal	0.48	0.48	0.49	0.49	0.50	0.51	0.50	0.50	0.50	0.49	0.48	0.47	0.46	0.46	0.48	0.50	0.50	0.50	0.48	0.48
Estonia	0.36	0.35	0.35	0.36	0.38	0.39	0.40	0.40	0.42	0.40	0.41	0.42	0.41	0.42	0.46	0.48	0.48	0.48	0.47	0.48
Lithuania	0.38	0.38	0.40	0.41	0.40	0.41	0.41	0.41	0.41	0.41	0.41	0.42	0.42	0.43	0.46	0.48	0.49	0.48	0.51	0.48
Latvia	0.31	0.32	0.34	0.35	0.36	0.37	0.39	0.38	0.38	0.37	0.38	0.38	0.38	0.39	0.41	0.44	0.46	0.46	0.44	0.47
Italy	0.54	0.53	0.51	0.50	0.51	0.51	0.50	0.49	0.49	0.47	0.47	0.47	0.45	0.46	0.48	0.48	0.49	0.48	0.46	0.46
Georgia	0.34	0.32	0.31	0.33	0.35	0.36	0.39	0.39	0.39	0.40	0.40	0.41	0.41	0.43	0.46	0.46	0.47	0.46	0.43	0.44
Albania	0.37	0.36	0.37	0.36	0.37	0.37	0.38	0.39	0.40	0.41	0.41	0.42	0.42	0.42	0.45	0.46	0.47	0.46	0.43	0.43
Croatia	0.39	0.39	0.39	0.39	0.41	0.42	0.42	0.42	0.42	0.41	0.42	0.42	0.42	0.42	0.43	0.44	0.44	0.46	0.41	0.43

Poland	0.38	0.38	0.39	0.38	0.38	0.39	0.38	0.38	0.40	0.40	0.42	0.42	0.41	0.41	0.44	0.44	0.44	0.43	0.41	0.42
Romania	0.38	0.37	0.36	0.36	0.36	0.38	0.38	0.38	0.38	0.37	0.39	0.38	0.38	0.40	0.42	0.42	0.42	0.42	0.40	0.41
Hungary	0.40	0.40	0.41	0.41	0.42	0.43	0.44	0.42	0.42	0.40	0.41	0.42	0.41	0.41	0.42	0.43	0.43	0.42	0.39	0.40
Greece	0.50	0.50	0.50	0.49	0.49	0.50	0.50	0.50	0.50	0.49	0.46	0.44	0.41	0.40	0.41	0.41	0.42	0.41	0.39	0.39
Serbia	0.31	0.32	0.35	0.34	0.36	0.35	0.37	0.37	0.37	0.36	0.37	0.37	0.37	0.39	0.42	0.42	0.42	0.41	0.39	0.39
Azerbaijan	0.30	0.29	0.30	0.30	0.30	0.30	0.30	0.30	0.30	0.31	0.31	0.32	0.31	0.35	0.37	0.40	0.41	0.39	0.37	0.37
Armenia	0.34	0.34	0.35	0.34	0.34	0.34	0.33	0.33	0.32	0.34	0.35	0.36	0.37	0.38	0.39	0.39	0.40	0.39	0.36	0.37
Moldova	0.34	0.33	0.34	0.34	0.33	0.36	0.35	0.35	0.34	0.33	0.33	0.35	0.36	0.37	0.40	0.41	0.41	0.40	0.37	0.37
Bulgaria	0.32	0.32	0.33	0.33	0.34	0.35	0.35	0.35	0.35	0.34	0.34	0.34	0.35	0.35	0.37	0.38	0.38	0.38	0.35	0.36
Bosnia and Herz.	0.33	0.33	0.33	0.34	0.36	0.35	0.36	0.35	0.36	0.35	0.36	0.35	0.38	0.38	0.39	0.39	0.41	0.39	0.35	0.36
Turkiye	0.32	0.32	0.32	0.32	0.33	0.36	0.36	0.36	0.36	0.35	0.35	0.36	0.36	0.36	0.38	0.37	0.36	0.36	0.33	0.33
Kazakhstan	0.27	0.26	0.26	0.26	0.26	0.27	0.27	0.26	0.27	0.29	0.28	0.28	0.29	0.30	0.33	0.35	0.35	0.34	0.32	0.32
Kyrgyz Republic	0.32	0.31	0.31	0.31	0.31	0.32	0.34	0.32	0.31	0.29	0.30	0.31	0.33	0.35	0.35	0.34	0.34	0.33	0.31	0.30
Ukraine	0.23	0.23	0.23	0.24	0.24	0.26	0.27	0.27	0.26	0.25	0.27	0.27	0.28	0.28	0.28	0.29	0.30	0.30	0.27	0.28
Russian Federation	0.21	0.21	0.23	0.21	0.21	0.23	0.24	0.24	0.23	0.24	0.22	0.23	0.24	0.26	0.28	0.28	0.28	0.28	0.26	0.27
Uzbekistan	0.21	0.21	0.22	0.21	0.21	0.21	0.22	0.22	0.22	0.21	0.21	0.22	0.23	0.24	0.26	0.27	0.27	0.26	0.23	0.24
East, South Asia, and the Pacific																				
Australia	0.68	0.68	0.68	0.69	0.69	0.70	0.70	0.71	0.72	0.71	0.73	0.72	0.72	0.70	0.72	0.70	0.73	0.73	0.74	0.74
Japan	0.63	0.62	0.61	0.63	0.64	0.65	0.66	0.66	0.66	0.64	0.67	0.66	0.67	0.66	0.68	0.66	0.69	0.69	0.69	0.69
New Zealand	0.57	0.58	0.59	0.59	0.59	0.60	0.61	0.62	0.63	0.63	0.65	0.65	0.66	0.64	0.66	0.65	0.68	0.68	0.68	0.69
Korea, Rep.	0.46	0.49	0.50	0.51	0.52	0.54	0.54	0.57	0.58	0.57	0.59	0.60	0.60	0.59	0.60	0.57	0.61	0.61	0.63	0.64
Malaysia	0.43	0.43	0.44	0.45	0.45	0.47	0.48	0.48	0.48	0.46	0.50	0.50	0.51	0.49	0.52	0.50	0.53	0.53	0.55	0.55
China	0.37	0.37	0.38	0.39	0.39	0.39	0.39	0.40	0.42	0.42	0.42	0.43	0.43	0.44	0.44	0.45	0.45	0.45	0.46	0.47
Vietnam	0.30	0.30	0.30	0.31	0.31	0.33	0.33	0.34	0.36	0.35	0.36	0.38	0.39	0.38	0.40	0.38	0.42	0.42	0.42	0.43
India	0.29	0.29	0.28	0.30	0.31	0.33	0.33	0.34	0.35	0.35	0.35	0.35	0.35	0.35	0.34	0.34	0.36	0.36	0.36	0.37
Thailand	0.33	0.33	0.34	0.34	0.32	0.32	0.31	0.31	0.32	0.31	0.31	0.33	0.34	0.32	0.32	0.32	0.33	0.33	0.33	0.35
Indonesia	0.23	0.22	0.21	0.22	0.22	0.22	0.24	0.26	0.28	0.27	0.28	0.29	0.30	0.28	0.30	0.27	0.32	0.32	0.34	0.35
Philippines	0.20	0.21	0.22	0.22	0.21	0.22	0.21	0.22	0.23	0.23	0.26	0.27	0.28	0.26	0.28	0.25	0.28	0.28	0.28	0.29
Pakistan	0.23	0.22	0.20	0.23	0.22	0.23	0.24	0.24	0.25	0.25	0.26	0.26	0.26	0.24	0.25	0.23	0.26	0.26	0.26	0.25

Source: author's estimates

A similar trend of growth inclusiveness is observed in East Asia, South Asia, and the Pacific (Table 5.2). Throughout the whole period, the region's four advanced economies—Australia, Japan, New Zealand, and the Republic of Korea—represent the leading four. In various years, Indonesia, Thailand, Pakistan, Philippines, and Thailand are in the trailing quartet in terms of inclusiveness. China is consistently in the middle ranks.

Considering that a key objective of our study is identifying causality between the inclusive growth of national economies and consumption of different types of energy, we proceed to analyzing the obtained results. Evaluation between IGI and energy consumption has been performed on the basis of econometric specifications (5) using R language. We will make the following remark. Implementing of three models and selection of the best of them is exclusively used to reveal the causality between inclusive growth of economy and energy consumption. Models are not used for subsequent forecasting or other purposes.

Econometric Analysis Empirical results of the three models (excluding nuclear energy) for Europe and CA as a whole are as follows. In all three models, five of the six indicators—specific CO_2 emissions, coal, gas, oil, and RES consumption—all demonstrate a statistically significant (at the $p < 0.01$ level) relationship between the response function (dependent variable IGI) and the predictors (energy consumption). Consequently, consumption of these energy sources in the Europe-Central Asia region explains the causal relationship with the 34-country inclusive growth index. However, the best statistical estimates (SE, R^2, Adjusted R^2, F-Statistic) are typical for model 3. In all three models, a positive impact on inclusiveness of oil, coal, and RES consumption growth but negative impacts of gas, hydroelectricity, and specific CO_2 emissions have been found. In contrast, coal consumption did not indicate the importance of the relationship with the dependent variable (Table 5.3).

Econometric model results for the East, South Asia, and Pacific region are as follows. First, for R^2, Adjusted R^2 and F-Statistic criteria, Model 2 is more convincing, while Model 3 is more convincing for SE. However, the determination coefficients in both models demonstrate that the quality of the models is relatively "not bad" ($0.8 < R^2 > 0.5$) (a model is very good if $R^2 > 0.8$). In Model 2, response function is positively influenced by predictors of oil, gas, coal and hydropower consumption, and negatively influenced by RES and specific CO_2 emissions. But for coal consumption only, a statistically significant relationship with a probability of 99% ($p < 0.01$) can be reported. Model 3 revealed a statistically significant positive effect of gas and coal consumption on the dependent variable, and a negative effect of oil, coal, hydroelectricity, RES, and specific CO_2 emissions on the dependent variable (all variables except hydroelectricity at the $p < 0.01$ level).

Considering that the analyzed regions include very different economies with varying development levels, different energy consumption patterns and other factors, subsequent study was performed by dividing the two groups of countries into subregions.

Table 5.3 Empirical estimation of the relationship between the evolution of the Inclusive Growth Index and energy consumption for Europe and Central Asia, East Asia, South Asia, and the Pacific

	Model 1: two effects		Model 2: individual		Model 3: time	
	Estimate	SE	Estimate	SE	Estimate	SE
Europe and Central Asia						
Log oil	0.21***	(0.02)	0.18***	(0.02)	0.03***	(0.01)
Log gas	−0.04***	(0.01)	−0.05***	(0.01)	−0.04***	(0.01)
Log coal	0.001	(0.001)	0.0004	(0.001)	0.0002	(0.001)
Log hydro	−0.02***	(0.01)	−0.02**	(0.01)	−0.03***	(0.002)
Log renew	0.01***	(0.002)	0.01***	(0.002)	0.03***	(0.003)
Log dummy_renew	−0.04***	(0.01)	−0.06***	(0.01)	−0.13***	(0.01)
Log CO_2	−0.10***	(0.02)	−0.14***	(0.01)	−0.21***	(0.01)
Observations	680		680		680	
R^2	0.33		0.55		0.87	
Adjusted R^2	0.27		0.52		0.87	
F-Statistic	43.70*** (df = 7; 620)		109.62*** (df = 7; 639)		646.67*** (df = 7; 653)	
East, South Asia, and Pacific						
Log oil	−0.01	(0.03)	0.02	(0.03)	−0.28***	(0.06)
Log gas	0.01	(0.01)	0.01	(0.01)	0.15***	(0.03)
Log coal	0.09***	(0.01)	0.13***	(0.01)	0.24***	(0.03)
Log hydro	0.02**	(0.01)	0.02	(0.01)	−0.02	(0.02)
Log renew	−0.01**	(0.003)	−0.004	(0.003)	−0.03***	(0.01)
Log dummy_renew	0.01	(0.01)	−0.02*	(0.01)	0.01	(0.05)
Log CO_2	−0.08*	(0.04)	−0.25***	(0.03)	−0.65***	(0.04)
Observations	240		240		240	
R^2	0.30		0.73		0.57	
Adjusted R^2	0.18		0.71		0.52	
F-Statistic	12.53*** (df = 7; 202)		86.45*** (df = 7; 221)		40.96*** (df = 7; 213)	

Source: own estimates

Note: Hereinafter SE, standard error of the coefficient; $*p < 0.1$; $**p < 0.05$; $***p < 0.01$, significance at the 10%, 5%, and 1% level, respectively

Subregions of Europe and CA First of all, it should be noted that the five subregions within the Europe and CA region differ markedly on the inclusive growth level. In particular, Western and Northern Europe are particularly affected by the highest average IGI values over the whole period (0.57 and 0.51 respectively), increasing in 2019 compared to 2000 by 5.3% and 16.8% respectively. South and East European subregions have relatively good IGI index (0.45 and 0.41 respectively), showing a growth of almost 14% in the southern subregion and a decline of 1.8% in the eastern subregion over two decades. Russia, Kazakhstan, and Uzbekistan have very low growth index inclusiveness: average IGI for the whole period is 0.28 with a growth rate of 20.1%.

Results of the empirical estimation of the relationship between growth inclusiveness, consumption of energy sources and specific CO_2 emissions by subregions of

Europe and CA are given in Table 5.4. Econometric models for northern Europe, including nine countries (Great Britain, Denmark, Ireland, Latvia, Lithuania, Norway, Finland, Sweden, and Estonia) have been implemented as follows. Model 1 is rejected due to low quality (R^2 value = 0.48). Model 2 in multiple coefficients of multiple determination is good and model 3 is very good. Model 2 shows a positive relationship of the endogenous variable (IGI) with the exogenous variables "oil" and "RES", and a negative relationship with the variables "gas" and "specific CO_2 emissions" at a high level of significance ($p < 0.01$). In Model 3, most predictors, except oil and gas, show high statistically significant effects on the response variable. Thus, coal consumption and RES have a positive impact on the inclusive growth of the North Europe economies.

For Eastern Europe, including nine countries (Azerbaijan, Bulgaria, Czech Republic, Hungary, Poland, Romania, Russia, Slovakia, and Ukraine), the best statistical estimates are specific to model 3. Thus, RES consumption has a stimulative effect on the variable of interest, while gas use and specific CO_2 emissions have a delaying effect at the significance level of $p < 0.01$. All other exogenous variables are statistically insignificant.

In Western Europe, including seven countries (Austria, Belgium, France, Germany, Luxembourg, Netherlands, and Switzerland) positive influence on the inclusiveness of national economies growth shows RES, and statistically significant negative influence—oil, hydropower, and specific CO_2 emissions at the significance level of $p < 0.01$, and coal at the level of $p < 0.1$ (model 3). Statistical estimates of gas consumption are untenable.

In South Europe, including six countries (Croatia, Greece, Italy, Portugal, Slovenia, and Spain) the second and third models are good. Estimates obtained for all exogenous variables except gas in model 2 are statistically relevant at the significance level of $p < 0.05$ and $p < 0.01$. Moreover, using oil, coal, and renewable energy catalyzes the inclusive growth of the subregion's economies. Conversely, negative predictors of gas consumption, hydroelectricity and CO_2 emissions are inhibitors of inclusive growth.

In Russia, Kazakhstan, and Uzbekistan, the best statistics significant relationship between endogenous and exogenous variables was revealed by model 3. In particular, a positive impact at the significance level of $p < 0.05$ was found from RES consumption and a negative impact from gas consumption at the significance level of $p < 0.01$. Significantly, model 3 estimated specific CO_2 emissions failed, and model 2 revealed a negative impact at a high level of significance ($p < 0.01$).

East, South Asia and Pacific Subregions Understandably, the two subregions that belong to the truncated Asia-Pacific region (APR) noticeably vary in inclusive growth. Thus, it is considerably higher in the East Asia and Pacific subregion, including the four developed economies (Australia, Japan, New Zealand, Korea, Rep.). Its average for the 20-year period was 0.6416 with a 17% increase. In the Southeast Asian subregion covering eight developing economies in our study (Malaysia, China, Vietnam, India, Thailand, Indonesia, Philippines, and Pakistan),

Table 5.4 Results of empirical estimation of the correlation between the inclusive growth index, energy consumption, and specific CO_2 emissions by subregions of Europe and Central Asia

	Model 1: two effects		Model 2: individual		Model 3: time	
Variables	Estimate	SE	Estimate	SE	Estimate	SE
Northern Europe						
Log oil	0.21***	(0.05)	0.18***	(0.04)	0.002	(0.02)
Log gas	−0.02	(0.02)	−0.06***	(0.02)	−0.02*	(0.01)
Log coal	−0.004	(0.01)	−0.01	(0.01)	0.03***	(0.01)
Log hydro	0.001	(0.01)	−0.003	(0.01)	−0.002	(0.002)
Log renew	0.02***	(0.005)	0.03***	(0.003)	0.04***	(0.01)
Log dummy_renew	−0.08***	(0.02)	−0.12***	(0.02)	−0.11***	(0.03)
Log CO_2	0.03	(0.03)	−0.07***	(0.02)	−0.14***	(0.02)
Observations	180		180		180	
R^2	0.48		0.75		0.90	
Adjusted R^2	0.35		0.73		0.88	
F Statistic	18.81*** (df = 7; 145)		71.07*** (df = 7; 164)		197.86*** (df = 7; 153)	
Eastern Europe						
Log oil	0.05	(0.05)	0.12***	(0.04)	−0.03*	(0.02)
Log gas	0.04	(0.03)	0.01	(0.03)	−0.06***	(0.01)
Log coal	−0.0002	(0.001)	−0.0002	(0.001)	0.0004	(0.001)
Log hydro	−0.07***	(0.02)	−0.06***	(0.02)	0.003	(0.01)
Log renew	−0.001	(0.003)	0.01***	(0.003)	0.03***	(0.01)
Log dummy_renew	−0.01	(0.01)	−0.06***	(0.01)	−0.16***	(0.02)
Log CO_2	0.01	(0.03)	−0.11***	(0.02)	−0.21***	(0.02)
Observations	180		180		180	
R^2	0.18		0.59		0.86	
Adjusted R^2	−0.01		0.55		0.84	
F Statistic	4.49*** (df = 7; 145)		33.56*** (df = 7; 164)		135.44*** (df = 7; 153)	
Western Europe						
Log oil	−0.04	(0.02)	−0.07*	(0.03)	−0.03***	(0.01)
Log gas	0.04*	(0.02)	0.0003	(0.02)	0.01	(0.01)
Log coal	0.02	(0.02)	0.04**	(0.02)	−0.02*	(0.01)
Log hydro	−0.06***	(0.01)	−0.03**	(0.01)	−0.02***	(0.002)
Log renew	0.03***	(0.01)	0.02***	(0.01)	0.02***	(0.01)
Log CO_2	0.11**	(0.04)	−0.08**	(0.04)	−0.03*	(0.02)
Observations	140		140		140	
R^2	0.35		0.67		0.89	
Adjusted R^2	0.16		0.63		0.87	
F Statistic	9.69*** (df = 6; 108)		42.30*** (df = 6; 127)		158.89*** (df = 6; 114)	
Southern Europe						
Log oil	0.35***	(0.04)	0.32***	(0.03)	−0.02	(0.02)
Log gas	−0.03	(0.02)	−0.09***	(0.02)	−0.08***	(0.02)
Log coal	0.03***	(0.01)	0.02**	(0.01)	0.06***	(0.01)
Log hydro	−0.06***	(0.02)	−0.04***	(0.01)	0.03	(0.02)
Log renew	0.02***	(0.01)	0.01**	(0.01)	−0.0003	(0.01)

(continued)

Table 5.4 (continued)

	Model 1: two effects		Model 2: individual		Model 3: time	
Variables	Estimate	SE	Estimate	SE	Estimate	SE
Log dummy_renew	-0.07^{***}	(0.02)	-0.06^{***}	(0.02)	−0.01	(0.03)
Log CO_2	-0.43^{***}	(0.08)	-0.18^{***}	(0.05)	-0.36^{***}	(0.07)
Observations	120		120		120	
R^2	0.71		0.61		0.67	
Adjusted R^2	0.61		0.57		0.58	
F Statistic	31.09^{***} (df = 7; 88)		24.33^{***} (df = 7; 107)		27.11^{***} (df = 7; 93)	
Russia, Kazakhstan, and Uzbekistan						
Log oil	−0.04	(0.07)	0.13	(0.09)	0.04	(0.06)
Log gas	0.11^{**}	(0.05)	0.06	(0.07)	-0.08^{***}	(0.02)
Log coal	-0.10^{**}	(0.04)	-0.11^{*}	(0.06)	0.03	(0.02)
Log hydro	−0.04	(0.06)	−0.02	(0.08)	−0.03	(0.05)
Log renew	−0.003	(0.01)	0.003	(0.01)	0.01^{**}	(0.01)
Log dummy_renew	0.01	(0.03)	−0.02	(0.04)	-0.07^{**}	(0.03)
Log CO_2	−0.05	(0.08)	-0.29^{***}	(0.05)	0.05	(0.08)
Observations	60		60		60	
R^2	0.30		0.76		0.95	
Adjusted R^2	−0.34		0.72		0.91	
F Statistic	1.86 (df = 7; 31)		22.61^{***} (df = 7; 50)		93.53^{***} (df = 7; 33)	

Source: own estimates

the similar indicators were 0.3349 and 28%. IGI changes presented indicate that the rate of growth of inclusiveness in the Southeast Asian subregion is relatively high.

In the developed economies of Asia and the Pacific, two models out of three are good. But the best statistical estimates at the 1% level with a high closeness of relationship ($R^2 > 0.9$) are characteristic of model 3: the positive influence of coal consumption, and the restraining influence of oil, hydropower and RES consumption and, indeed, CO_2 emissions. In the Southeast Asian subregion, model 2 is good: a positive impact on inclusiveness is revealed for increases in coal and hydropower consumption at the 1% and 10% significance levels, respectively, and a negative impact for RES consumption and specific CO_2 emissions at the 5% and 1% significance levels, respectively (Table 5.5).

Econometric specifications for Europe and Asian countries using nuclear energy were constructed in the study (Table 5.6). In Europe and Asia, with 17 and 5 countries using nuclear power respectively, all sources have a statistically significant impact on the inclusive growth of their economies. It is noteworthy that in both regions, nuclear energy consumption has a negative impact on inclusive growth dynamics. The following reasons seem to be the reason for this to a certain extent. Despite an 18.7% increase in nuclear consumption in Asia & Pacific in 2019 compared to 2000, its share in Primary Energy declined by 49% and in Europe & CA both absolute volumes and share declined by 14%.

Table 5.5 Results of empirical estimation of the correlation between the inclusive growth index, energy consumption and specific CO_2 emissions by subregions of East, South Asia and the Pacific subregions

	Model 1: two effects		Model 2: individual		Model 3: time	
Variables	Estimate	SE	Estimate	SE	Estimate	SE
East Asia and Pacific						
Log Oil	−0.03	(0.04)	−0.03	(0.03)	-0.15^{***}	(0.01)
Log Gas	0.01	(0.03)	0.01	(0.03)	0.01	(0.02)
Log Coal	0.13^{***}	(0.03)	0.12^{***}	(0.03)	0.18^{***}	(0.01)
Log Hydro	0.02	(0.03)	0.02	(0.02)	-0.04^{***}	(0.01)
Log Renew	0.004	(0.01)	0.01	(0.01)	−0.001	(0.01)
Log CO_2	-0.37^{***}	(0.10)	-0.33^{***}	(0.05)	-0.61^{***}	(0.08)
Observations	80		80		80	
R^2	0.55		0.84		0.94	
Adjusted R^2	0.30		0.82		0.92	
F Statistic	10.37^{***} (df = 6; 51)		63.55^{***} (df = 6; 70)		153.04^{***} (df = 6; 54)	
South-East Asia						
Log Oil	−0.03	(0.05)	0.04	(0.04)	-0.20^{**}	(0.08)
Log Gas	0.01	(0.01)	0.01	(0.01)	0.04	(0.02)
Log Coal	0.07^{**}	(0.03)	0.13^{***}	(0.01)	0.12^{***}	(0.04)
Log Hydro	0.04^{**}	(0.02)	0.03^{*}	(0.02)	-0.07^{***}	(0.02)
Log Renew	-0.01^{***}	(0.004)	-0.01^{**}	(0.004)	0.01	(0.01)
Log dummy_renew	0.03	(0.02)	0.01	(0.02)	0.003	(0.05)
Log CO2	−0.01	(0.07)	-0.17^{***}	(0.04)	0.61^{***}	(0.09)
Observations	160		160		160	
R^2	0.24		0.75		0.51	
Adjusted R^2	0.04		0.72		0.42	
F Statistic	5.70^{***} (df = 7; 126)		61.54^{***} (df = 7; 145)		20.01^{***} (df = 7; 133)	

Source: own estimates
Note: SE, standard error of the coefficient; $*p < 0.1$; $**p < 0.05$; $***p < 0.01$, significance at the 10%, 5% and 1% level, respectively

Thus, implementation of the econometric models allows for intermediate conclusion. Firstly, the main hypothesis of the study on the impact of energy consumption on the inclusiveness of growth of national economies is confirmed. Second, according to Model 3, in Europe and CA, all energy resources except coal have multidirectional effects on the growth inclusiveness of its economies, and in East, South Asia and Pacific, except hydropower. It is impossible to explain the multidirectional influence solely by constructed and implemented econometric dependencies, as growth inclusiveness is a complex phenomenon and depends on many factors. This is a subject for future research. Third, RES consumption is a driver of inclusive growth in Europe and Central Asia and a constraint in the Asia-Pacific region. It is possible that this is predetermined by the consumption structure of energy sources. Finally, and clearly, in the two regions and their subregions, with the exception of the three CA countries (Model 3), the constraint to inclusive growth of national economies is specific CO_2 emissions (for all good models).

Table 5.6 Results of empirical estimation of the relationship between the inclusive growth index, energy consumption and specific CO_2 emissions for European and Asian countries using nuclear energy

	Model 1: two effects		Model 2: individual		Model 3: time	
Variables	Estimate	SE	Estimate	SE	Estimate	SE
European countries (including nuclear energy consumption)						
Log oil	0.09***	(0.02)	0.06**	(0.03)	−0.06***	(0.01)
Log gas	0.002	(0.01)	−0.01	(0.01)	−0.02***	(0.01)
Log coal	0.03***	(0.01)	0.04***	(0.01)	0.09***	(0.01)
Log hydro	−0.02*	(0.01)	−0.02	(0.01)	−0.03***	(0.003)
Log renew	0.002	(0.002)	0.002	(0.003)	0.01***	(0.004)
Log dummy_renew	−0.003	(0.01)	0.002	(0.01)	−0.04*	(0.02)
Log CO_2	−0.11***	(0.03)	−0.21***	(0.02)	−0.34***	(0.01)
Log nuclear	0.01	(0.01)	−0.01	(0.01)	−0.01*	(0.01)
Observations	340		340		340	
R^2	0.15		0.59		0.93	
Adjusted R^2	0.02		0.56		0.93	
F Statistic	6.32*** (df = 8; 296)		57.19*** (df = 8; 315)		558.13*** (df = 8; 312)	
Asian countries (including nuclear energy consumption)						
Log oil	0.13***	(0.03)	0.05	(0.03)	0.16***	(0.03)
Log Gas	−0.04*	(0.02)	0.07***	(0.02)	−0.01	(0.02)
Log coal	0.07***	(0.02)	0.10***	(0.02)	0.12***	(0.02)
Log hydro	−0.03*	(0.02)	−0.05***	(0.02)	−0.08***	(0.01)
Log renew	0.001	(0.002)	−0.003	(0.003)	−0.01***	(0.002)
Log dummy_renew	0.01	(0.01)	0.01	(0.01)	0.04**	(0.01)
Log CO_2	−0.35***	(0.07)	−0.19***	(0.06)	−0.54***	(0.02)
Log nuclear	−0.02***	(0.01)	−0.01*	(0.01)	−0.02***	(0.01)
Log dummy_Nuclear	0.01	(0.01)	0.01	(0.02)	0.02	(0.02)
Observations	100		100		100	
R^2	0.46		0.80		0.99	
Adjusted R^2	0.20		0.77		0.99	
F Statistic	6.30*** (df = 9; 67)		39.26*** (df = 9; 86)		1299.77*** (df = 9; 71)	

Source: own estimates

Note: SE, standard error of the coefficient; *$p < 0.1$; **$p < 0.05$; ***$p < 0.01$, significance at the 10%, 5% and 1% level, respectively

5.5 CO_2 Consumption and Emissions: Structural and Regional Analysis

An analysis of energy consumption patterns over the period 2000–2021 shows that oil, gas and coal account for more than 80% of energy consumption in Europe and CA, with oil and gas accounting for an average of 65% of energy consumption over these years. In the analyzed period, the shares of consumption decreased: oil from 34% to 29% and coal from 18% to 13%. Notably, at the beginning of the pandemic

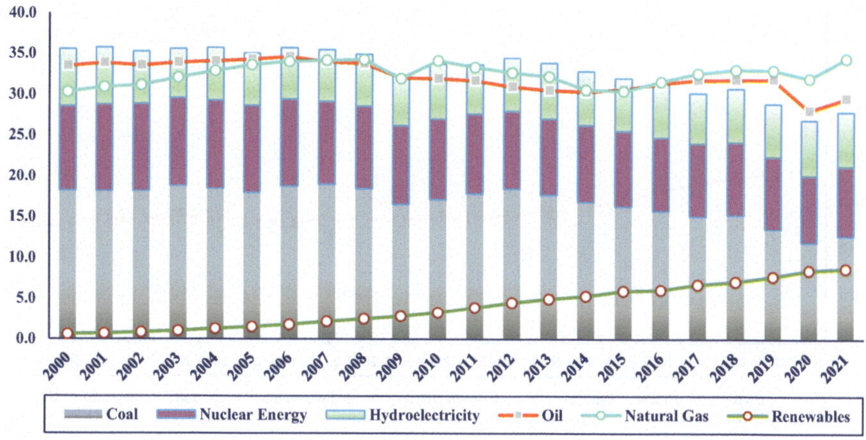

Fig. 5.2 Dynamics of change in energy consumption shares in the Europe and CA region (as % of primary energy), 2000–2021. Source: calculated and compiled by the authors according to BP Statistical Review World Energy (2022). Note: axis X—years; axis Y—percentage

period (2020), the shares of oil, gas, coal and nuclear energy consumption declined markedly, reaching the lowest value for oil, coal and nuclear energy over the entire 22-year period. The combined share of the three fossil energy sources (oil, gas and coal) reached a low point of 72% in 2020 (78% in 2019). Share of hydropower consumption remained stable, while the use of RES showed significant growth, increasing more than 14 times (Fig. 5.2).

It is widely known that the largest share of CO_2 emissions comes from coal combustion, with a global share of 40% in 2021, followed by oil (32%) and gas (21%). CO_2 emissions from the Europe-Central Asia region as a share of global absolute CO_2 emissions decreased significantly over the past 22 years, from 27% to 16%.[5] In pandemic 2020, total CO_2 emissions in the 34 countries analyzed are reduced by 10%. In 2021, the impact of the pandemic was almost negated: CO_2 emissions increased by 6%. However, in absolute volumes, they have not reached 2019 levels. Top three anti-leaders in terms of emissions are (in descending order): Russia, Germany and Turkey. On the contrary, the outsiders in CO_2 pollution are (in ascending order) Latvia, Luxembourg and Slovenia.

During the period 2000–2021, energy consumption in the East, South Asia and Pacific region follows the following trends and structural shifts. Coal is the main source of energy in the region, accounting for an average of 51% over the entire period. Over the years, its share grew from 44% to 48%, and in the crisis and post-crisis years 2008–2011 it was between 54% and 55%. Oil is the second most important and consumed energy source, although its share has declined from 37% to 25% in more than two decades. The third significant fossil energy source is gas. Its share rose insignificantly, from 9% to 10%. The aggregate share of oil, gas and coal

[5] Source: here and below estimated by the authors from data (bp Review, 2022).

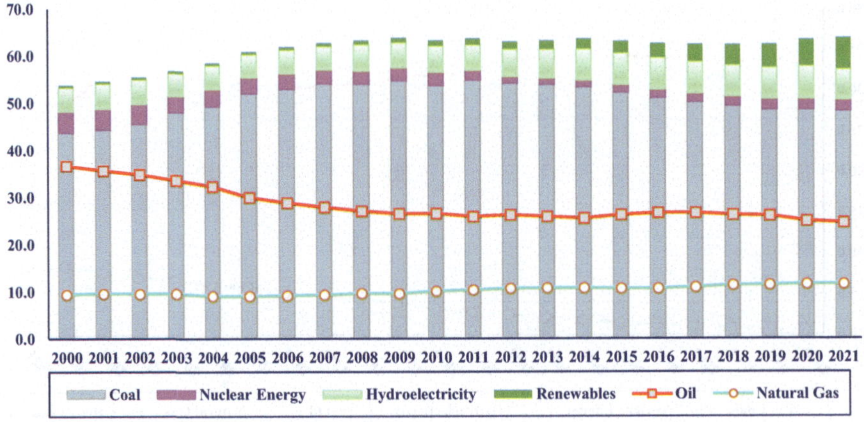

Fig. 5.3 Dynamics of change in energy consumption shares in East, South Asia and Pacific (as % of primary energy), 2000–2021. Source: calculated and compiled by the authors according to BP Statistical Review World Energy (2022). Note: axis X—years; axis Y—percentage

consumption fell from 90% to 84% over a 22-year period. Over the years, the use of nuclear energy has been reduced by two and a half times, from 5% to 2%. The use of RES increased significantly, from 1% to 7% (Fig. 5.3).

Europe and CA Subregions (Charts of changes in energy consumption by all subregions are presented in Appendix B) In the Nordic countries, the trend of increasing the use of non-conventional energy sources is clearly visible. Thus, there is a rapid growth of RES consumption, the share of which increased from 2% to 18% between 2000 and 2021, increasing in absolute volumes by 9.4 times. Use of coal has steadily declined, with its share falling by almost two-thirds over the years, from 13% to 4%. Nuclear power consumption is also declining, driven by a marked decrease in its use in the UK and the closure of the Ignalina nuclear power plant (NPP) in Lithuania in 2009. During the analyzed period, there was a decrease in the share of consumption of the other two types of fossil energy sources: oil by 9%, gas by 8%.

Among the Nordic countries, the UK has the highest CO_2 emissions, with an average share of 64% of the total. It's understandable that this is due to the size and power of the British economy. However, within 22 years, emissions have decreased by 40%. Minimum emissions are recorded in 2020–2021. In total, CO_2 emissions decreased by 36% in the subregion. Emissions have been reduced by more than half in Denmark, by more than a third in Finland, which is second in volume, and by 30% in Sweden, which is second only to the UK in volume. With the exception of the Baltic States, where CO_2 emissions are on the rise, all countries in the subregion have reduced their emissions. The highest growth of emissions (12%) is characteristic for Lithuania.

In Eastern Europe, the share of total fossil energy consumption fell from 88% to 85% in the new millennium. The analysis shows that this trend is solely due to a decrease in the share of coal utilization (from 25% to 16%), caused by a 34% decrease in absolute volumes. At the same time, there was an increase in the share of oil utilization (from 20% to 23%) and gas utilization (from 44% to 46%). The use of nuclear energy increased by 17%, including a twofold increase in the Czech Republic and one and a half times in Russia. It can be stated that the subregion has made a bias towards the use of RES, the share of which is still relatively small, but has increased 63 times in 22 years, from 0.03% to 2.0%. Poland is the leader in RES utilization in the subregion, followed by Romania.

Total CO_2 emissions in Eastern Europe decreased by almost 4% in the new millennium. Ukraine is the leader in terms of emission reduction rates (by 50%), followed by the Czech Republic (by 26%) and Romania (by 21%). Azerbaijan is the anti-leader in terms of emissions growth rate (24%), followed by Russia (by 9%) and Poland (by 3%). Certainly, the maximum volumes of emissions come from Russia, whose share against the background of their cumulative decline over the years has increased from 59% to 67%. At the beginning of the noughties Ukraine ranked second in terms of emissions, but given the significant reduction in emissions, by the end of the period it had moved to third position, passing Poland.

In Western Europe, the main energy sources are fuels (oil, gas and coal), with a combined average share of 72% over two decades, of which oil 38%, gas 21% and coal 13%. At the same time, the dynamics and share of oil and coal utilization is decreasing, while gas utilization is increasing. Germany is the main gas consumer in the subregion, with an average over the whole period of 43%. The volume and share of nuclear energy consumption is declining markedly, primarily due to a nearly triple decline in Germany. The volumes and share of hydropower utilization are slightly decreasing, while RES are increasing.

In this subregion, the main polluter is Germany, which emits more than half of all CO_2 emissions (52% on average over the period). This is largely due to the fact that it consumes almost ¾ of all coal used in Western Europe. It is followed by France with 23%, then the Netherlands with 14%. Collectively, these three countries emit 88% of the CO_2 of the entire subregion. However, in all 7 countries of Western Europe, the dynamics of CO_2 emissions shows a decrease, except for Luxembourg, where carbon dioxide emissions have increased by 3% over 22 years. It is remarkable what is happening in Luxembourg at the same time as fossil energy consumption is declining.

In Southern Europe, oil prevails among energy sources, and the rate of consumption and share in the resource mix have decreased markedly over 22 years: dynamics—by a third, share—by a quarter. The main consumers of black gold in the subregion are Italy and Spain. The second most utilized energy source is gas, the share of which increased 1.5 times and the rate 1.4 times. Italy and Spain are also leaders in the use of this type of energy resource. In Italy, the dynamics of blue fuel use remains relatively stable (7% growth over the whole period), while in Spain there is an increasing dynamic of volumes of its consumption (1.9 times). The largest increase in gas consumption is observed in Slovenia (5.5 times), followed by Greece

(3.6 times) and Portugal (2.5 times). Total gas utilization in the subregion increased by 36%. Against this background, the share of coal consumption decreased by 71% and the share of hydropower decreased by 9%. At the same time, there was a significant increase of RES in all countries of Southern Europe without exception, which in aggregate was reflected in a 14-fold increase of its share in primary energy.

In relation to carbon dioxide emissions, the main polluters are Italy and Spain, which reduced their CO_2 emissions by 28% and 21% respectively in the analyzed period. A decrease in CO_2 emissions is characteristic of all countries in the subregion. Thus, Greece reduced emissions by 45%, Portugal—by 35%, and the lowest rate of reduction of carbon dioxide emissions is noted in Croatia. The total CO_2 emissions in the subregion decreased by 28%.

In the smallest subregion by the selected number of countries (Russia, Kazakhstan and Uzbekistan), a characteristic trend is the multidirectional dynamics of primary energy consumption: in Kazakhstan there is a 2.1 times growth, in Russia—by 21%, and in Uzbekistan a 9% decrease. The absolute dominant energy resource is natural gas, the share of which consumption averaged 52% over the entire period. In aggregate, fossil resources (oil, gas and coal) account for 89% of primary energy, with consumption increasing by 21% over the entire period. Virtually all types of energy consumption, with the exception of coal (down 3%), have increased in the subregion over the years. It is remarkable that the consumption of nuclear energy has increased one and a half times. The share of RES consumption in primary energy is very small—0.04% on average over the whole period.

CO_2 emission volumes in two countries of the subregion are increasing: in Kazakhstan—by 2.2 times, in Russia—by 9%. In Uzbekistan, this indicator has decreased by 8% over 22 years. More than twofold growth of CO_2 emissions in Kazakhstan is associated with the same growth (2.1 times) in coal consumption.

East, South Asia and Pacific Subregions In East Asia and Pacific countries, the combined share of oil, gas and coal consumption averaged 86% over 22 years, with black gold dominating at 43%. However, the latter's demand share has fallen from 49% to 39% over the years. At the same time, the share of consumption of the other two fossil sources increased: gas from 12% to 20% and coal from 22% to 26%. Over the years, the share of nuclear energy consumption has declined significantly, from 12% to 5%. This trend was the result of a significant decrease in consumption of this energy source in Japan by 83%, while South Korea's consumption grew by 28%. As a result, total nuclear energy consumption in the subregion decreased by 55%. Over the years, there has been a tenfold increase in the volume and share of renewable energy consumption, with a slight decrease in the use of hydropower.

In the analyzed subregion, the largest volume of CO_2 emissions falls on Japan, whose share in total emissions averaged 55% over 22 years, with a 14% decrease over the entire period. The highest dynamics of CO_2 emissions growth is observed in South Korea (by 38%), where the consumption of fossil fuels has increased significantly over the years: oil—by 22%, natural gas—by 3.2 times, coal—by 1.7 times. In Australia and New Zealand, the growth in carbon dioxide emissions is relatively low, at 4% and just over 1%, respectively.

The following trends are evident in Southeast Asia. Firstly, the main source of energy is coal, with an average share of 58% over the whole period. Moreover, its consumption grew rapidly in all the countries of the subregion, by a factor of 3 in total. But particularly high growth rates are characteristic for Vietnam (11 times). It is followed by: Malaysia (almost 9 times) and Pakistan (8 times). In China and India, where for obvious reasons (multi-billion population) coal use is very significant, the rate of growth in each has tripled. Secondly, the use of natural gas has demonstrated a fourfold increase over the years, although its share in primary energy averages 9%. The growth of oil consumption volumes increased 2.4 times, with an average share of 24%. Third, nuclear energy consumption is characterized by rapid growth—12.4 times. Finally, RES utilization in the subregion has shown double-digit growth rates since 2004. Such growth dynamics, naturally, affected the increase in the share of RES, rising from 0.4% to 6.6% in primary energy. There is also an increase in the share of hydropower utilization, even though at a slower rate.

The dominant use of coal as an energy source is certainly reflected in the CO_2 emission limits. Thus, within 22 years, the subregion's share of global carbon dioxide emissions increased from 21% to 44%, with a threefold increase in volumes. It is clear that China has the largest volume and share of CO_2 emissions. Its share of global emission volumes increased from 14% to 31% in 22 years at a growth rate of more than 3%. The second country in the subregion that actively emits carbon dioxide into the atmosphere is India (a 2.7-fold increase). Among the countries of the subregion, the highest growth rate of CO_2 emissions is characteristic of Vietnam—5.7 times. Note that all 8 countries in the subregion included in the study show a trend of continuous multiple growth in carbon dioxide emissions.

Therefore, the structural analysis of energy sources utilization in the studied subregions allowed us to identify certain trends. Firstly, the dominance of fossil fuel use—oil, natural gas and coal—is evident in all subregions. Their average share over 22 years ranges from 70% in Northern Europe to 91% in Southeast Asia, and the average for all subregions is 82%. In particular, in terms of the share of demand for black gold in the consumption of energy resources the leaders are: Southern Europe (47%), East Asia and Pacific (43%). It is remarkable that Russia and CA have the lowest average consumption share of all subregions (21%).

Secondly, natural gas is the main energy source in Russia and CA, as well as in Eastern Europe, with a 52 and 44% share in total consumption, respectively. In the other three subregions of Europe, their share ranges between 21 and 26%. The Asia-Pacific region, especially Southeast Asia, has the least gas consumption (9%).

Third, Southeast Asia is the main consumer of coal, accounting for 58% of all 6 types of energy resources used in the subregion. It is followed by East Asia and Pacific at 26%, then Eastern Europe at 21%, Russia and CA at 16%. The lowest shares of coal use are in Northern Europe (11%) and Southern Europe (10%).

Fourth, the share of RES utilization is low in all subregions, but Northern and Southern Europe are the leaders (7% each), followed by Western Europe (6%). In Russia and Central Asia RES is practically not used. This indicates a significant perspective potential for the development of green energy. The situation is very similar in the utilization of hydropower. But here the undisputed favorite is Northern

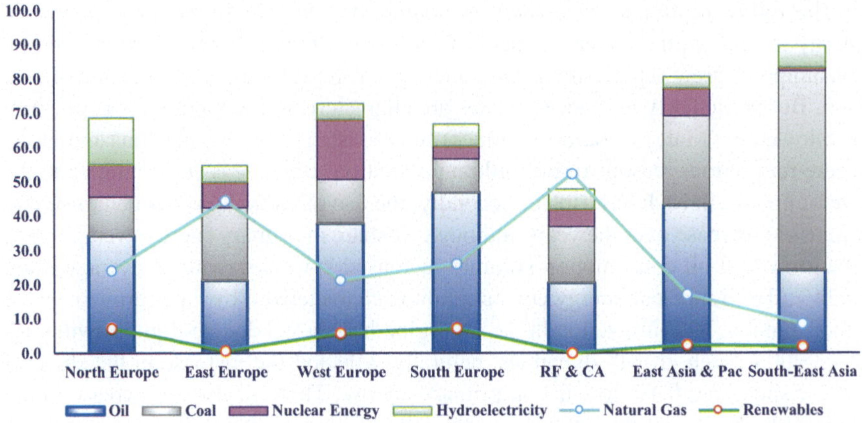

Fig. 5.4 Dynamics of change in energy consumption shares in subregions (as % of primary energy), 2000–2021 (average values for the period). Source: calculated and compiled by the authors according to BP Statistical Review World Energy (2022). Note: axis X—subregions; axis Y—percentage

Europe, which accounts for 14% of consumption from the primary energy subregion. In the remaining subregions, the share of hydropower use in the energy structure varies between 3 and 6% (Fig. 5.4).

Fifth, there is a clear downward trend in the share of nuclear energy in the energy balance of the European and Asian regions. Thus, even as its consumption in Southeast Asia grew by 32%, its share in the subregion declined by 47% over 22 years. Four countries in the subregion, excluding Japan, have increased their use of nuclear energy: China 22 times, Pakistan 15 times, India 2.5 times and South Korea 1.3 times. In contrast, Europe and CA showed declines in both speed and share of consumption by 17%. The maximum share of nuclear energy use in total energy consumption is observed in Western Europe—18%. Next are: Northern Europe—10%, East Asia and Eastern Europe—8% each (Fig. 5.5).

The specifics and structural features of energy consumption in the two studied regions have certainly affected the volume and rate of change of CO_2 emissions. In Europe and CA, the aggregate share of emissions has almost decreased twofold over the past more than two decades, from 52% to 29%. In contrast, in South, East Asia and the Pacific, the rate increased substantially, from 49% to 71%. This dynamic growth of CO_2 emissions in the region is due to the strong emission that occurs in Southeast Asia, whose share has grown 1.8 times (Table 5.7). European regions, with the exception of Russia and Central Asia (up 14%), have seen a permanent decrease in carbon dioxide emissions. In East Asia and Pacific, emissions in 2021 remained at 2000 levels, while in Southeast Asia they nearly tripled.

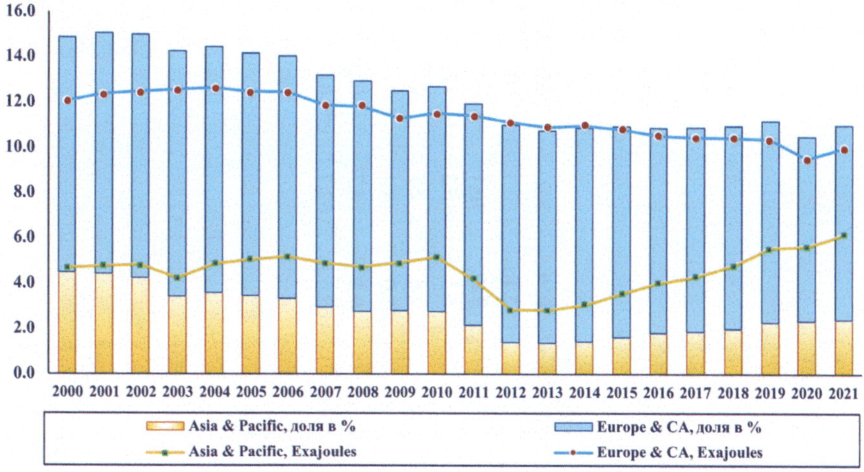

Fig. 5.5 Dynamics of change in rates and shares (as % of primary energy) of nuclear energy consumption in Asia and Europe, 2000–2021. Source: calculated and compiled by the authors according to BP Statistical Review World Energy (2022). Note: axis X—subregions; axis Y—percentage

5.6 Conclusion

In recent years, academic research has increasingly focused on the problem of effectively utilizing socioeconomic and environmental potential to increasingly meet the needs of the population. In this context, the present work is a fundamentally new study and aims to contribute to the explanation of the phenomenon of "inclusive growth" and its causal relationship with the new energy paradigm. The cascade of information on these two complex phenomena and processes and their causality can (and does) generate cognitive distortions due to inadequate interpretation of them. The obvious unevenness and ambiguity of achieving carbon neutrality by the countries of the world will surely increase inequality not only globally, but also at the intercountry and national levels. Global challenges and fractures in the architectonics of international relations, which occurred after February 2022, significantly increase the need for serious developments in this direction. Our substantive analysis of a plethora of publications reveals few research papers that explore the causality between inclusive growth and consumption of different energy sources across regions and countries of the world.

The novelty of this paper is also expressed in the fact that, unlike other studies, the Inclusive Growth Indices (IGI) for 55 Eurasian and Asian countries are constructed for a relatively long, 20-year period, which is predetermined by the purpose and objectives of the study. Therefore, the main criterion for selecting variables for determining the IGI was exclusively the availability of completeness of qualitative data for their dynamic series. It should be noted that the highest values of IGI are

Table 5.7 CO_2 emission shares in subregions, 2000–2021 (percentage of total)

Subregions	2000	2005	2010	2015	2016	2017	2018	2019	2020	2021	2021/2000
Northern Europe	5.8	4.8	4.0	3.1	3.0	2.9	2.8	2.6	2.4	2.3	40.0
Eastern Europe	16.8	13.6	11.8	10.5	10.6	10.4	10.4	10.1	9.9	10.1	60.0
Western Europe	11.2	9.0	7.5	6.5	6.5	6.4	6.1	5.8	5.3	5.2	47.0
Southern Europe	6.4	5.8	4.2	3.5	3.4	3.4	3.4	3.2	2.8	2.9	45.1
Russia and CA	11.4	9.4	8.6	8.2	8.2	8.1	8.2	8.1	8.0	8.1	71.0
East Asia and Pacific	14.0	12.1	10.7	10.3	10.1	10.0	9.7	9.5	9.1	8.7	62.4
South-East Asia	34.5	45.4	53.1	58.0	58.2	58.9	59.4	60.7	62.5	62.7	181.8

Source: calculated and compiled by the authors from bp Statistical Review (2022)

found in the developed economies of two regions. The developed countries of Europe and the Asia-Pacific region are ahead of the underdeveloped economies of the regions, being in the top half of the list ranked by IGI value during the two analyzed decades. Undoubtedly, the level of socioeconomic development, quality of life and a number of other key indicators of the advanced countries of these regions are much higher, as well as the distribution of benefits from growth, compared to the developing economies of Europe and the Asia-Pacific region.

The realization of econometric models allows us to state the following. First, the main hypothesis of the study about the impact of energy consumption on the inclusiveness of growth of national economies is confirmed. Secondly, in the region "Europe and Central Asia" all energy resources, except for coal, have a multidirectional impact on the inclusive growth of their economies, and in East, South Asia and Pacific—with the exception of hydropower. It is not possible to explain the multidirectional effects solely by econometric relationships, as growth inclusiveness is a complex phenomenon that depends on many factors and is the subject of future research. Thirdly, RES consumption is a stimulating factor of growth inclusiveness in Europe and Central Asia and a restraining factor in the Asia-Pacific region. Finally, and obviously, in all countries except Kazakhstan, Russia and Uzbekistan, specific CO_2 emissions are a constraint to inclusive growth of national economies.

Structural analysis of the use of energy sources in the studied subregions has revealed certain trends. First, in all subregions there is an obvious dominance of fossil fuel use—oil, natural gas and coal. In terms of the share of demand for "black" gold in the consumption of energy resources, two subregions are leading: Southern Europe, as well as East Asia and Pacific. Second, in Russia and Central Asia, as well as Eastern Europe, natural gas is the prevailing energy source. Third, Southeast Asia is the main consumer of coal. The smallest shares of coal use are in Northern and Southern Europe. Fourth, the share of RES utilization is insignificant in all subregions. However, the leading positions are occupied by three subregions: Northern, Southern and Western Europe. In Russia and Central Asia RES is practically not used. Theoretically, this indicates a significant promising potential for the development of "green" energy in post-Soviet countries. A very similar situation is observed in the use of hydropower. But here Northern Europe is the undisputed favorite. In other subregions, the share of hydropower use in the energy consumption structure is insignificant. Fifthly, there is a clear tendency for the share of nuclear energy use in the energy balance of the European and Asian regions to decrease. Thus, even with a 32% growth of its consumption in South-East Asia, its share in the subregion has almost halved between 2000 and 2021. In contrast, Europe and Central Asia have experienced a nearly 20% decline in both the rate and share of nuclear energy consumption.

Undoubtedly, the declared new paradigm of "Future Energy" will entail an unprecedented expansion and transformation of the global energy sector infrastructure. This will require a massive expansion of renewable energy and energy efficiency to meet rapidly growing demand. The envisaged decarbonization of the energy sector in countries around the world implies: firstly, a phase-out of coal-fired power generation, the main source of CO_2 emissions; secondly, RES generation must accelerate significantly. However, to finance the energy transition, the world's countries will have to mobilize much more capital than they do today. For example, the World Bank (2023) estimates that low- and middle-income countries will need up to US$1 trillion by 2030 for energy transition-related activities. There are significant challenges to the realism of achieving carbon neutrality by 2050–2060 in developed economies as well. This is especially true for European countries. After 2022, in the context of the energy crisis, many states made adjustments to their plans, primarily related to moving too quickly away from fossil fuels. Many developed countries have stated that their strategies to phase out coal-fired power plants are unchanged. In reality, the risk of an energy famine in practice is forcing them to consider more conservatively projects to switch from coal and other fossilized natural energy resources to renewable sources.

Finally, and most importantly. We dare to argue that this paper is at the forefront of a number of pilot studies to assess and identify causal links between inclusive growth and the new energy paradigm. The obtained results of econometric models and structural analysis of energy consumption by regions and subregions of the world convince us of the need for further research. By supplementing the econometric toolkit with more complex economic and mathematical models, it is possible to obtain multidimensional results. First, it will be possible to carry out scenario calculations to assess the impact of the energy consumption structure of different countries and regions of the world on their inclusive growth in the medium and long term. Secondly, scenario calculations will make it possible to assess the realities of the achievability of carbon neutrality in specific countries and regions.

Acknowledgments This research is funded by the Science Committee of the Ministry of Science and Higher Education of the Republic of Kazakhstan (Grant No. BR21882364).

References

Aizenman, J., Lee, M., & Park, D. (2012). The relationship between structural change and inequality: A conceptual overview with special reference to developing Asia. In *ADBI Working Paper 396*. Asian Development Bank Institute. Available: http://www.adbi.org/working-paper/2012/11/13/5332.structural.change.inequality.dev.asia/

Atkinson, E. (2018). *Inequality: What can be done?* (Russ. ed.: Neravenstvo: kak s nim byt'?). Transl. from. Eng. by O. Levchenko. Moscow: Delo. 536 p. (In Russ.).

Economic Growth (Russ. Ed.: Ekonomicheskij rost) Transl. from. Eng. by O. Kapusti-na, A. Moiseev. Moscow: BINOM. 2010. 824 p. (In Russ.).

Borovsky, Y. V. (2021). Energy security problem amid global energy transition. *Vestnik RUDN. International Relations, 21*(4), 772–784. (In Russian). https://doi.org/10.22363/2313-0660-2021-21-4-772-784

bp Statistical Review of World Energy: 71st Edition. (2022). https://www.bp.com/en/global/corporate/energy-economics/statistical-review-of-world-energy.html

Chancel, L., Piketty, T., Saez, E., Zucman, G., et al. (2022). *World Inequality Report 2022*. World Inequality Lab. wir2022.wid.world.

Che Sulaiman, N. F., Ab-Hamid, M. F., & Ridzuan, A. R. (2021). Inclusive growth: Comprehensive dimension of income distribution. *Economy of Region, 17*(1), 301–317. https://doi.org/10.17059/econ.reg.2021-1-23

Commission on Growth and Development. (2008). *The growth report: Strategies for sustained growth and inclusive development*. Washington, DC: World Bank. http://hdl.handle.net/10986/6507

Cui, L., Weng, S., & Song, M. (2022). Financial inclusion, renewable energy consumption, and inclusive growth: cross-country evidence. *Energy Efficiency, 15*(6), 43. https://doi.org/10.1007/s12053-022-10051-y

Dent, C. M. (2012). Renewable energy and East Asia's new developmentalism: towards a low carbon future? *The Pacific Review, 25*(5), 561–587. https://doi.org/10.1080/09512748.2012.728240

Dharmaratne, M. A., & Attygalle, M. D. T. (2018). Improving the statistical capacity index: A statistical approach. *American Journal of Applied Mathematics and Statistics, 6*(3), 80–95. https://doi.org/10.12691/ajams-6-3-1

Dong, J., Liu, D., Dou, X., Li, B., Lv, S., Jiang, Y., & Ma, T. (2021). Key issues and technical applications in the study of power markets as the system adapts to the new power system in China. *Sustainability, 13*, 13409. https://doi.org/10.3390/su132313409

EUROPE 2020: A European Strategy for Smart, Sustainable and Inclusive Growth. (2010). European Commission.

Goldberg, P. K., & Reed, T. (2023). Is the global economy deglobalizing? And if so, why? And what is next? National Bureau of Economic Research, Working Paper 31115.

Grigoryev, L., Makarov, I., Sokolova, A., Pavlyushina, V., & Stepanov, I. (2020). Climate change and inequality: How to solve these problems jointly? *International Organisations Research Journal, 15*(1), 7–30. (in English). https://doi.org/10.17323/1996-7845-2020-01-01

Heshmati, A., Kim, J., & Wood, J. (2019). A survey of inclusive growth policy. *Economies, 7*, 65. https://doi.org/10.3390/economies7030065

Huo, T., Xu, L., Feng, W., Cai, W., & Liu, B. (2021). Dynamic scenario simulations of carbon emission peak in China's city-scale urban residential building sector through 2050. *Energy Policy, 159*, 112612. https://doi.org/10.1016/j.enpol.2021.112612

Khusainov, B. D., Kaimoldina, S. A., & Nussupov, A. I. (2021). Growth quality and income inequality: Cross-country analysis. *Living Standards of the Population in the Regions of Russia, 17*(4), 454–461. https://doi.org/10.19181/lsprr.2021.17.4.3

Khusainov, B., Nussupov, A., Kaimoldina, S., & Shirov, A. (2022). Assessment of the quality of growth of national economies in the context of digital transformation. In *Digital Transformation and the World Economy: Critical Factors and Sector-Focused Mathematical Models* (pp. 67–86). Springer.

Lukash, S. (2021). G20 in the second decade of the 21st century. Steering the world towards inclusive growth and sustainable development. *International Organisations Research Journal, 16*(2), 7–14. (in English). https://doi.org/10.17323/1996-7845-2021-02-01

Milanovich, B. (2022). The three eras of global inequality, 1820-2020 with the focus on the past thirty years. In *Stone Center Working Paper Series, No 59* (p. 52). Stone Center on Socio-Economic Inequality.

OECD. (2008). *Handbook on constructing composite indicators. Methodology and user guide* (p. 158). OECD Publishing. www.oecd.org/publishing

OECD. (2018a). *Opportunities for all: A framework for policy action on inclusive growth* (p. 200). OECD Publishing. https://doi.org/10.1787/9789264301665-en

OECD. (2018b). *The productivity-inclusiveness nexus*. OECD Publishing. https://doi.org/10.1787/9789264292932-en

Parra-Domínguez, J., Sánchez, E., & Ordóñez, Á. (2023). The prosumer: A systematic review of the new paradigm in energy and sustainable development. *Sustainability, 15*(13), 44. https://doi.org/10.3390/su151310552

Patcharaporn, L., Chiara, C., & Tomohide, M. (2023). Macroeconomic shocks and conflict. *IMF Working Paper, 2023*(068), 42. https://doi.org/10.5089/9798400238222.001

Piketty, T. (2018). *Capital in the twenty-first century* (Russ. Ed.: Kapital v XXI veke). Transl. from. Eng. by A. Dunaev. AD Marginem Press. P. 592. (In Russ.).

Porfiriev, B. N., & Shirov, A. A. (2022). Strategies for socioeconomic development with low greenhouse gas emissions: Scenarios and realities for Russia. *Herald of the Russian Academy of Sciences, 92*(3), 230–238. https://doi.org/10.1134/S1019331622030030

Rytova, E., Gutman, S., & Sousa, C. (2021). Regional inclusive development: An assessment of Russian Regions. *Sustainability, 2021*(13), 5773. https://doi.org/10.3390/su13115773

Sharafutdinov, R. I., Akhmetshin, E. M., Polyakova, A. G., Gerasimov, V. O., Shpakova, R. N., & Mikhailova, M. V. (2019). Inclusive growth: A dataset on key and institutional foundations for inclusive development of Russian Regions. *Journal Data in Brief, 23*, 103864. https://doi.org/10.1016/j.dib.2019.103864

Smorgunov, L. V. (2018). Inclusive growth and administrative reform in the brics countries. *Public Administration Issues*, Special Issue (electronic edition), pp. 80–95 (in English). https://doi.org/10.17323/1999-5431-2018-0-5-80-95.

Stawska, J., & Jabłon'ska, M. (2022). Determinants of inclusive growth in the context of the theory of sustainable finance in the European Union countries. *Sustainability, 14*, 100. https://doi.org/10.3390/su14010100

Stiglitz, J. (2015). *The price of inequality: How today's divided society endangers our future* (Russ. ed.: Cena neravenstva. Chem rassloenie obshchestva grozit nashemu budushchemu). Transl. from. Eng. by E. Rozhdestvenskaya. Eksmo. P. 508. (In Russ.).

Stiglitz, J. (2020). *People, power, and Profits: Progressive capitalism for an age of discontent* (Russ. ed.: Lyudi, vlast' i pribyl': Progressivnyj kapitalizm v epohu massovogo nedovol'stva). Transl. from. Eng. by V. Ionov. Al'pi-na Pablisher. P. 430. (In Russ.).

Stiglitz, J., Sen, A., & Fitoussi, J.-P. (2009). Report by the commission on the measurement of economic performance and social progress.

Strategy of achieving carbon neutrality of the Republic of Kazakhstan until 2060. (2023). Decree of the President of the Republic of Kazakhstan from February 2, 2023, №121. https://adilet.zan.kz/rus/docs/U2300000121

Tsapko-Piddubna, O. (2021). Inclusive growth policy and institutional assessment: The case of Central and Eastern European Countries. *Baltic Journal of Economic Study, 7*(2), 233–239. https://doi.org/10.30525/2256-0742/2021-7-2-233-239

UNCTAD. (2018). Inclusive Growth of the Eurasian Economic Union Member States: assessments and opportunities.

Vlahinić Lenz, N., & Fajdetić, B. (2021). Globalization and GHG emissions in the EU: Do we need a new development paradigm? *Sustainability, 13*, 9936. https://doi.org/10.3390/su13179936

WEF. (2018). The inclusive development index 2018: http://www3.weforum.org/docs/WEF_Forum_IncGrwth_2018.pdf

World Bank. (2023). *Scaling up to phase down: Financing energy transition in the power sector.* World Bank.

World in Data CO2 and Greenhouse Gas Emissions dataset. (2022). https://github.com/owid/co2-data#data-on-co2-and-greenhouse-gas-emissions-by-our-world-in-data

Yakovlev, I. A., Kabir, L. S., Nikulina, S. I., & Rakov, I. D. (2020). The impact of the sustainable development agenda on the transformation of national policies of commodity producing countries. *Ekonomika regiona (Economy of Region), 16*(3), 859–870. https://doi.org/10.17059/ekon.reg.2020-3-14

Popкova, E., Sergi, B., Haabazoka, L., & Ragulina, J. (2020). Supporting Inclusive Growth and Sustainable Development in Africa. In *Volume II: Sustainability in Infrastructure Development.* Springer Nature.

Chapter 6
Solutions to the Problems of Transition to Green Energy in Kazakhstan

Nickolay Buktukov, Yevgeniy Gumennikov, and Gulnaz Moldabayeva

Abstract Renewable energy sources such as wind farms and solar panels are technological solutions aimed at mitigating the greenhouse effect caused by modern industry including the coal industry. Coal energy brings about significant environmental damage by occupying vast areas of agricultural land for waste dumps and ash landfills. Opponents of renewable energy sources consider the use of installations working on wind and solar energy to be unacceptable due to a number of reasons: the variability of wind and solar energy; the use of very rare and expensive metals; the impossibility to recycle used blades of wind turbines and photocells of solar panels; the unreasonably high energy consumption at the production of these installations and higher cost of the generated energy compared to traditional energy sources. The chapter shows that there are solutions to all of these problems through the use of more complex energy generation technologies, including those in the coal industry.

6.1 Introduction

In the last 20 years, renewable energy sources (RES) have been rapidly developing, and the most widespread are wind and solar energy. At the same time, any new direction in human life has both supporters and opponents. In this regard, the use of renewable energy sources is no exception. The main arguments of renewable energy supporters are ecological problems associated with environmental degradation and growing carbon dioxide emissions, and as a result, climate change due to the widespread use of fuel energy, which creates a greenhouse effect.

N. Buktukov (✉) · Y. Gumennikov
The Mining Institute after D.A. Kunayev of the National Center for Complex Processing of Minerals of the Republic of Kazakhstan, Almaty, Kazakhstan

G. Moldabayeva
The Kazakh National Research University after K.I. Satpayev, Almaty, Kazakhstan
e-mail: g.moldabayeva@satbayev.university

© The Author(s), under exclusive license to Springer Nature Switzerland AG 2024
T. Devezas et al. (eds.), *Global Energy Transition and Sustainable Development Challenges, Vol. 1*, World-Systems Evolution and Global Futures,
https://doi.org/10.1007/978-3-031-67583-6_6

Opponents of the wide use of RES base their reasoning on what they think are insurmountable obstacles or unacceptable consequences, namely:

- The variability of wind and solar energy
- The use of very rare and expensive metals
- The impossibility to recycle used blades of wind turbines and photocells of solar panels at the current stage of development
- High energy consumption at the production of installations generating renewable energy
- Higher cost of the generated energy compared to traditional energy sources

One of the most aggressive sources of carbon dioxide emissions is coal, which indeed still plays an extremely important role in the energy balance of countries such as China and India. Kazakhstan is no exception. About 75% of electricity is generated by coal burning. However, the governments of many countries, including Kazakhstan, are taking measures to reduce carbon dioxide emissions. For example, the National Carbon Quota Plan for 2022–2025 plans to reduce greenhouse gas emission quotas from 151.8 million tons to 125.3 million tons in total for the electric power, oil and gas, mining, metallurgical, chemical, and construction material industries. Later in the chapter, we will take a closer look at specific technological solutions proposed to reduce carbon dioxide emissions both through the use of renewable energy sources and the modernization of the coal industry.

6.2 The Issue of Variability of Wind and Solar Energy

As is well known, wind is not everywhere and not always, but the sun is everywhere, but not always. In this case, it is possible to overcome the variability of energy sources, and it is already being done by combining solar panels and wind farms, which results in wind solar installations, as well as by using accumulators. This combination is widely used for consumer-grade RES, but has not yet gained ground for industrial-scale wind farms where energy is transferred to the grid.

In the first case, it is very important that the cost of accumulators is significant, especially of traction ones, where rare metals are used. The use of lead-acid accumulators is limited by a short service life, since it has fewer charge and discharge cycles. The solution lies in the use of energy storage devices, for example, those designed by Baeshov A.B., the member of the National Academy of Sciences of the Republic of Kazakhstan . The principle of the design is based on the use of one or more accumulators to charge a special liquid with a large electrical capacity. This liquid gives a charge back through the same accumulators when electricity is needed.

This principle of energy storage has been known for 138 years (Bradley, 1885) and currently gains grounds due to its low cost and less environmental hazard compared to modern accumulators and a number of other advantages.

A number of works are aimed at solving the problem of energy storage. The economic and environmental advantages of iron-air batteries (developed by Form

Energy) also contribute to its development. Developments by "The Batteries" are very gratifying since they offer the opportunity to accumulate energy, are durable, fast charging, and easily recyclable.

Another possible option is a capacitor-based energy storage device, which should be discharged in the same way as accumulators.

There are proposals in which pressurized air is accumulated and released to rotate the wind farm and generate electricity.

The simplest and most widely used energy storage option is hydraulic accumulators. Excess energy during the operation of wind, solar or wind-solar installations is used to pump water into a tank, and if there is no wind or sun, the water is released from the tank onto a hydroturbine that generates electricity.

Recently, gravity energy accumulators have begun to develop, the principle of which is based on the use of gravity forces. Excess energy from renewable energy sources is used to lift loads, and when there is a need in electrical energy, the load is lowered, and the accumulated potential energy is converted into electrical energy. In this situation, there is no need in large areas compared to traditional accumulators. Very significant energy capacity can be accumulated on gravity accumulators when using depleted mines and quarries. For instance, in 2023–2025, "Gravitricity" plans to create a 4 MW gravity accumulator. It should be noted that the service life of such accumulators is at least 50 years, and there are no problems with their utilization.

In mountainous terrain, gravity energy accumulators can be successfully used on mountain slopes.

In addition, biogas can be used to generate heat and/or electric energy.

All these additional costs, except for the costs of biogas plants, must be compensated when wind and solar energy is used. And this is possible without a significant increase in the cost of energy if wind farms generate energy in the entire range of wind speeds, and if solar panels operate with high efficiency and are of low cost. This can be achieved and will be discussed below.

Thus, the operation of wind farms in a wide range of wind speeds is possible with an automatic increase or decrease in the swept area. In addition, it must be taken into account that at low wind speeds the most effective are sail blades, and at medium wind speeds, aerodynamic blades. In this regard, Fig. 6.1 (Buktukov et al., 2019) shows a wind power plant that operates effectively in the wind speed range from 5–7 up to 50 m/s.

Figure 6.1 shows that the wind farm has sail blades in the form of torus segments, and aerodynamic blades, the design of which is made in such a way that at low wind speeds they work like sail blades. Sail blades are wide open if there is no wind and at low wind speeds due to the load connected to the blade through a flexible traction. The aerodynamic blades are connected to the sail blades through a flexible traction and are somewhat turned inward. This and the very design of the blades ensure their operation like sail blades at low wind speeds.

At medium wind speeds, the sail blades partly close down due to the wind pressure and transfer the aerodynamic blades to their normal position, and they work as aerodynamic ones. In a strong wind, the sail blades close almost completely also due to the wind pressure and continue operating, and the aerodynamic blades go

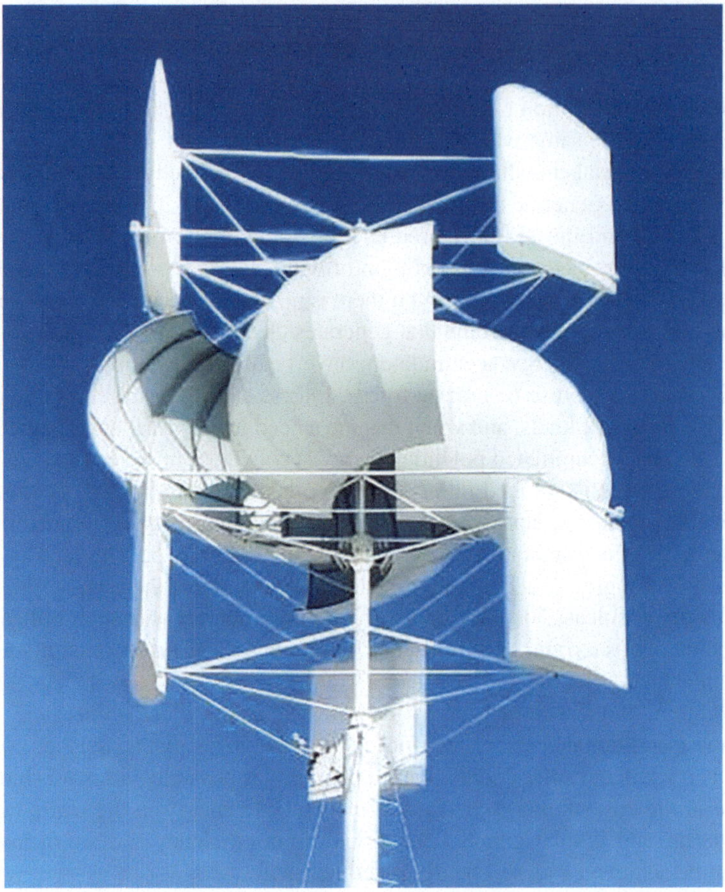

Fig. 6.1 Wind farm (a photo of the production prototype)

into braking mode. This ensures ongoing operation and the number of revolutions, as experiments have shown, does not increase, remaining within 20 revolutions per minute. The braking mode is performed by deflecting the aerodynamic blades outward relative to their shaft.

The operability of such a wind farm design does not depend on the direction and gusts of wind, it ensures the increase in the time utilization factors and rated power by more than two times without a noticeable decrease in efficiency. Consequently, the amount of generated energy also more than doubles, which makes it possible to use the listed types of energy storage units without the increase or with a slight increase in the energy cost.

The problem of variability of sun energy is handled in the same way as described above, and also by significantly increasing the efficiency of solar panels and reducing the number of expensive photocells by many times, which will be shown below at the discussion of the problem of rare metal deficit.

In addition to the above-mentioned methods of energy storage, research is being done around the world on the production of hydrogen and its subsequent use. Given the high energy consumption required for that, it is planned to use energy from renewable sources. In Kazakhstan similar research has been discontinued more then 30 years ago, although encouraging results were obtained before.

6.3 The Issue of Deficit of Rare and Expensive Metals

The problem of the limited use of rare metals at the first stage can be solved by their significant reduction. When using wind energy, rare metals are used in accumulators and generators in combination with wind power plants as well as when using solar panels. In this case, the use of rare metals in accumulators will be reduced at the creation of the above-mentioned energy storage devices. And it is also possible to reduce their use in generators almost twice (Fig. 6.2) (Buktukov et al., 2011).

Figure 6.2 illustrates that the drive shaft 1 rotates the drive bevel gear 2, which in turn rotates the driven bevel gears 3 and 4 in opposite directions relative to each other. Accordingly, shafts 5 and 6 rotate disks with winding and magnets in opposite directions. Therefore, the intersection of magnetic lines and windings is doubled compared to the rotation of only magnets or windings, as is customary in conventional generators. It makes it clear that with the counter rotation of magnets and windings according to Fig. 6.2, at the same capacity as a traditional generator, the need for rare and non-ferrous metals is twice as much less.

However, the large amount of rare metals is required in solar panels. Currently, the main direction in the development of solar panels is increasing the efficiency of photocells and keeping their cost low.

However, it is possible to aim at a significant increase in the efficiency of solar panels rather than photocells by dispersing solar radiation by wavelengths and focusing them on highly efficient photocells along the main optical axis. For

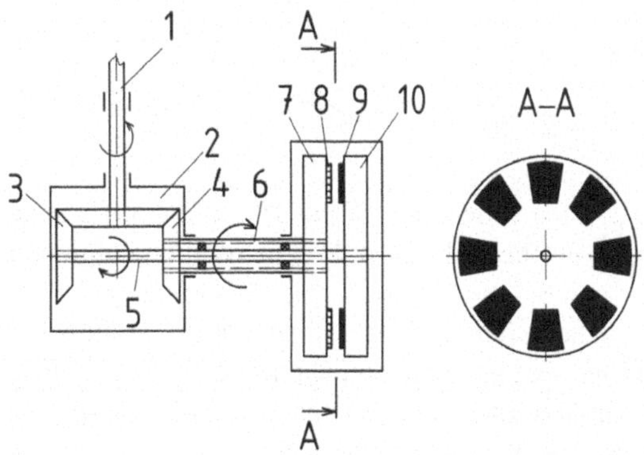

1 – the driving shaft of the wind farm; 2 – the drive bevel gear; 3, 4 – the driven bevel gears; 5 – the shaft connecting the driven bevel gear 3 with the disk 10, where the magnets 9 are installed; 6 – the shaft connecting the driven bevel gear 4 with the disk 7, where the windings 8 are installed.

Fig. 6.2 Generator

a)

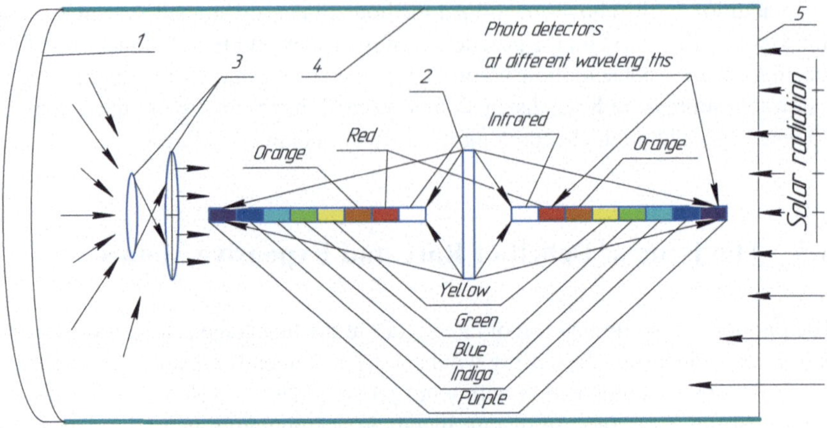

b)

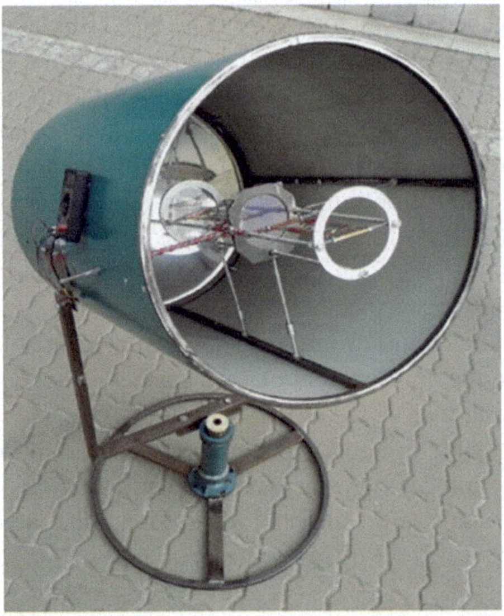

a – Solar panel schematic; b – A photo of the production prototype

Fig. 6.3 Solar-voltaic array. (**a**) Solar panel schematic; (**b**) A photo of the production prototype

example, Fig. 6.3 (Buktukov et al., 2016) shows one of the 18 options for a solar panel where solar radiation, falling on a spherical mirror 1, is focused on collimator lenses 3, which direct the rays in parallel to a holographic concentrator 2. The holographic concentrator 2 disperses the radiation by the wavelengths, as shown in Fig. 6.3, along the main optical axis and focuses them on the photocells. Some of the incoming solar radiation falls on the holographic concentrator 2 and is also dispersed and focused on the main optical axis and falls on the photocells.

In this case, it is possible to install an appropriate high efficiency photocell at a certain wavelength, even a very expensive one, since it is required in much smaller numbers and with lower parameters.

The efficiency of the solar battery as a whole and of the photocells increases not only because it receives the appropriate wavelength, but also due to the exclusion of infrared radiation from reaching the photocells. In this case, separately focused infrared radiation can be used to generate thermal energy by installing a metal receiver at the focus, and by placing its other end in a heat-removing liquid. In addition, a special photocell can be installed (Kiptilii & Gusinskii, 1987), which would convert it into electricity.

The experimental studies have shown the efficiency and feasibility of such a design of the solar panel. In this case, for example, provided a spherical mirror diameter is 60 cm and there are ten 5 × 5 mm photocells, the use of rare metals will be reduced by 1130 times. If the input flow of solar radiation is, for example, 1 m^2, the need in rare metals will be reduced by 4000 times. The limiting factor is the permissible amount of solar energy directed at the holographic concentrator.

As experiments have shown, in this regard there arises a need to establish the permissible number of suns directed at the holographic concentrator as well as the rational forms and parameters of photocells. If the photocells are made concave, it is advisable to install them behind the focal point, and convex ones, in front of the focal point. Then the sun's rays will fall almost perpendicular to the plane of the photocells. Manufacturing curved solar cells is not problematic (Vaidya & Solgaard, 2022; Saravanapavanantham, Mwaura, Bulović, 2022).

In any case, as experimental studies have shown, ways to reduce the use of rare metals while increasing the power and efficiency of solar photovoltaic panels have already been found.

It should be noted that the proposed design of solar panels does not require the use of significant areas of land (see below table), since they can be installed on wind power towers and poles.

Parameters and characteristics	"NEOSAN energy" solar panels	Proposed solar panels
Efficiency, %	Up to 20	45–60
For example, at 10 kW		
Cross-sectional area of solar radiation falling on the panel, m^2	63.5 (poly) 60.6 (mono)	20.2
The area of land required for the installation of a solar panel, m^2	70 (poly) 67 (mono)	1.5
Utilized capacity from 1 m^2, W	157.6 (poly) 165 (mono)	Up to 495

Another factor perceived as a deterrent to the development of wind solar energy is the problem of recycling of used blades and photocells.

6.4 The Issue of Recycling of Used Blades of Wind Farms and Photocells of Solar Panels at the Current Stage of Development

Since the foregoing and similar designs of wind farms, for example, by IceWind, Ehmberg Solutions and many other companies, are made without cantilever blade mounting, there is no need in very light and strong materials, which are very expensive and cannot be recycled at present. The blades and all other components of the proposed and similar designs are made or can be made of metal, which ensures their easy recycling.

As shown above, while the need in photovoltaic solar panels is reduced by thousands of times, their efficiency is significantly increased. In addition, new photocells are being developed (Bikbaev et al., 2022; Saravanapavanantham et al., 2022) not from rare metals, which can be easily used in the above-mentioned options of solar photovoltaic cells. Therefore, practically, there will be no problems with the disposal of photocells.

6.5 The Issue of Energy Consumption at the Production of Renewable Energy Sources

There is an opinion that the production of renewable energy sources results in a huge carbon emission, which does not justify their use. However, everything is comparative. For example, the manufacture of equipment for traditional energy sources—thermal and nuclear power plants—causes an even greater carbon emission. Moreover, fuel power plants continue causing emissions during their operation, including greenhouse gas, and nuclear power plants cause thermal pollution many times more than fuel ones, not counting the danger of radiation pollution.

It should be noted that carbon emissions from the production of renewable energy sources are compared with the operating life of wind power plants and solar panels, which is about 15–20 years. However, as experiments have shown, low-speed wind power plants and solar panels that separate infrared heat rays more than double their service life. Consequently, the amount of energy they produce is also much greater.

It is evident that this argument is not valid for refusing RES.

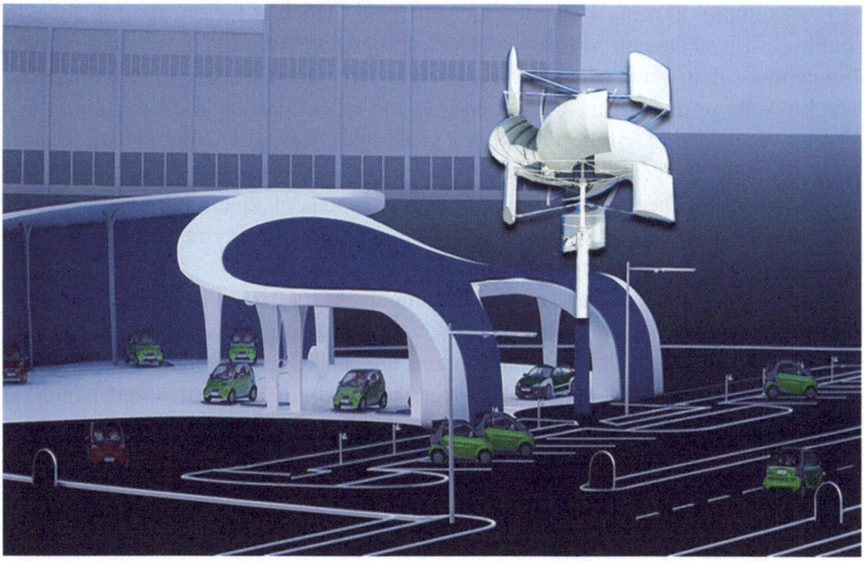

Fig. 6.4 A wind power plant near a charge station

6.6 The Issue of Higher Cost of Generated Energy Compared to Traditional Energy Sources

It is worth noting that in addition to the proposed designs of wind solar installations there are already other wind and solar installations, the cost of electricity from which approaches or does not exceed the cost of energy from electric grids, and which can often be installed not far from consumers, thereby reducing energy transportation costs. For instance, the wind power plant described above and its analogues do not produce any low-frequency or acoustic vibrations and can be installed near consumers (Fig. 6.4).

In addition, the cost of electricity from fuel and nuclear power plants does not include some significant costs. For example, the cost of a nuclear power plant does not take into account the costs of disposal, which is done during 2–3 decades, when the production of electricity has already been stopped. And it does not include security costs. In addition, the costs caused by the environmental damage in the form of emissions of ash, gases, radioactive and thermal pollution are not taken into account.

Similarly, the environmental damage from fuel thermal and power plants is not taken into account. In addition to greenhouse gas emissions, the coal energy, which in most cases is currently more efficient than nuclear power plants, wind farms and solar panels, poses a significant burden on the environment due to coal mining and transportation.

For example, to mine 1 ton of Ekibastuz coal, which is the cheapest in the CIS countries, it is necessary to load on vehicles, transport from the quarry and put into dumps more than 1 m^3 of rock. And as the quarry deepens, its amount increases, which results in the rise in the fuel cost. In the recent past, coal production at this deposit reached 115 million tons per year, and overburden amounted to over 120 million m^3 per year. Wastes from coal energy in the form of ash and overburden are laid on the surface of agricultural lands, cover tens of square kilometers and are blown away by the wind polluting the environment. Transportation of coal, overburden and ash also generate dust.

In particular, only one Ekibastuz plant annually burns 16 million tons of coal with an ash content of 47%, therefore, about seven million tons of ash are placed in ash dumps, and a little less than one million tons are carried away through the pipe. The ash contains sulfur, which is spread over thousands of kilometers and begins to oxidize the soil as soon as moisture comes into contact with it.

6.7 Underground Coal Reactor Is a Solution to Environmental and Economic Problems of Coal-Fired Energy

The idea of underground gasification was first put forward in 1888 (Mendeleev, n. d.), and in 1928 a project for underground gasification was developed (Bakulev, 1955). However, due to the low efficiency of technologies, but most importantly due to the discovery of large gas fields, this idea gradually declined. In addition, horizontal coal deposits were subjected to gasification, which could and were mined out very effectively by traditional methods.

At present, there are already scientific and technical developments that make it possible to burn coal in-situ and obtain thermal and electrical energy without polluting the environment at steep deposits that cannot be effectively worked out, and the purified greenhouse gas can be bubbled with alkali resulting in chalk and water (Buktukov, Gumennikov; 2022). To implement the new technology, special

Rectified current voltage – 53V;
Actual power consumption – 140kW;
Work-cycle time – 8 sec;
Preset pressure (for shutt-off valve) – 150 MPa;
Energy consumed during a cycle – 0.18 kW/hr;
Volume of broken rock with hardness scale 4 – 2,300 cm^3.
Unit weight – 322 kg.

Fig. 6.5 The production prototype of the hydraulic impulse sinking machine

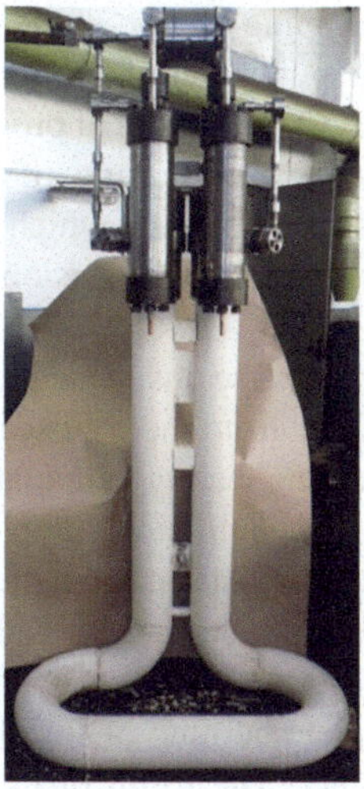

Compressor's drive steam pressure – 50 MPa;
Water column oscillation frequency – 4 oscillations per second;
50 MPa steam consumption per second – 0.0014 m^3/sec or 1.4 l/sec;
Energy consumption of an electric discharge drive of the steam generator per second – 254 kJ;
Weight of inert water in the compressor – 80 kg;
Performance of one U-shaped element at free air – 101.6 l/sec, or 6.1 m^3/min;
Gauge pressure of compressed air – 96 kg/cm^2;
Compressed air temperature – 635 C^0;
Compressor dry weight – 857 kg.

Fig. 6.6 The production prototype of the hydraulic percussion compressor

Impeller's diameter – 500 mm;
Working steam pressure – 50 MPa;
50 MPa steam consumption per second – 4.2 l/sec;
Average speed of the drive's water jet – 74.8 m/sec;
Speed of the loaded impeller – 1,510 rev/min.
Turbine's power output – 340 kW;
Energy efficiency – 57%.
Turbine weight without hydraulic amplifiers – 220 kg.

Fig. 6.7 The production prototype of the steam-hydraulic single-impeller turbine

equipment has been developed (now at the stage of operating models) for the construction of gasification sites and the conversion of thermal energy into electrical energy at power plants with a hydraulic turbine drive. In addition, ash and slag combustion products remain underground, and exhaust gases and suspensions are subjected to repeated purification during heating of neighboring gasification sites.

The proposed technology for the development of a steeply dipping coal deposit using underground combustion (gasification) has a number of positive aspects, in particular, a high degree of process control, easy and high-quality backfilling, minimization of surface disturbances, the absence of waste rock dumps and ash landfills and of coal transportation to thermal power plants, etc.

The process is the following. Initially, a steeply dipping section of the coal seam is prepared using (process) blast injection holes with a diameter of 300 mm or more by means of supersonic ejections of hydraulic charges to the bottom of process wells. The process wells are driven along the strike of coal seams, as a rule, of a complex and heterogeneous structure using a controlled tool string (Fig. 6.5) which is capable of non-contact destruction of both the coal mass and rock inclusions without deviations from the given direction.

Further, a steam-air blast is done at steeply dipping coal seams, which ensures the displacement of 10 MPa high-pressure formation water from the coal seam section mined using the fire method and, at the same time, a high heating temperature of the coal mass. To this end, a fundamentally new hydraulic percussion compressor was developed. It features one-stage compression (Fig. 6.6) with a very high pressure of a hot steam-air blast mixture in order to displace formation water outside the fire reactor and to activate gasification.

Process charts have been developed for the complete burning of steeply dipping seams to produce hot gas at the steam generator inlet. The result of this is high-temperature inert combustion products with a high compression density, the level of which is characterized by the thickness and gas impermeability of the overburden above the upper edges of the mined steeply dipping seams. The superheated steam gets onto the hydraulic percussion compressors and steam-hydraulic turbines (Fig. 6.7) and is converted into electrical energy. If the production is located near consumers, it is possible to supply water steam or hot water.

The basis of the new technology is the process of high-temperature high-pressure steam-air blast, which will be performed by means of hydrodynamic compressors of single-stage adiabatic compression, as shown above.

The blast pressure can reach 10 or more MPa and is capable of displacing formation water in steeply dipping coal seams at a depth of 800 m or more. Accordingly, the gasification process will be implemented at a gas pressure of up to 10 MPa.

In case of producing high-quality gas products, this implies a significant percentage of the generated methane and hydrogen, and in case of producing thermal energy, intensive fuel combustion in a high-ash environment with a high temperature of combustion products, which are used directly on industrial sites in heat-exchange steam generators with subsequent conversion of the kinetic energy of steam into electrical energy.

Development of steeply dipping coal seams with blast at a pressure of 10 MPa can give new better results in terms of the quality of resulting gasification products. For comparison, at blast pressures of 2–2.5 MPa in autoclave processes, in addition to CO and H_2, up to 15% of CH_4 is obtained (Geologiya, 1973).

The process charts below show the development of a blast injection hole, the installation of a downhole steam generator, the preparation of underground gasification sections, surface sections of underground gasification, an experimental section of underground gasification with the processing of thermal energy into electrical energy, and an example of underground gasification in relation to a specific field.

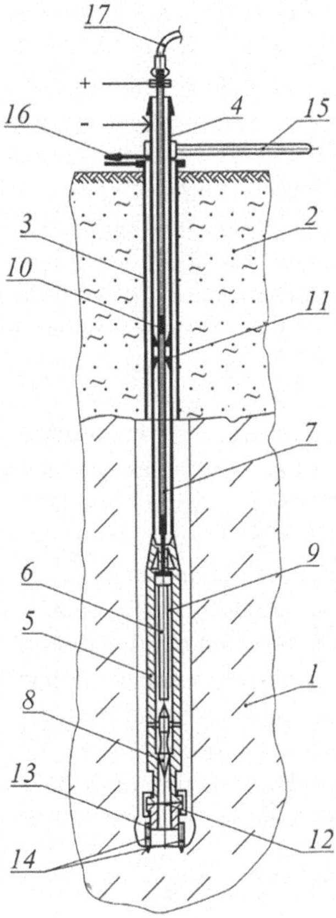

Fig. 6.8 Schematic view of driving a blast well with a hydraulic impulse coal sinking machine

The hydraulic impulse sinking machine (Fig. 6.8) is designed to prepare steeply dipping predominantly high-ash and watered coal seams 1. In addition, the wells from the surface along the overburden 2 can be opened by conventional rotary drilling and cased with steel pipes 3. Near the coal seam, the rotary string changes to a more flexible and rotating string 4 only in one fixed sector of 180°.

The flexibility of a semi-rotary drill string is increased due to a decrease in its diameter. This is possible because the method of drilling a well with a hydraulic impulse sinker does not require significant axial force.

At the downhole end of the string 4, an electric discharge hydraulic impulse unit 5 is mounted. Sector rotation of the string is used only to change the direction of tracing a driven well.

In the overburden of small and medium hardness, both vertical and continuing them process blast wells along the coal seam can be driven using a non-rotating method from the surface by means of an electric discharge hydraulic impulse unit 5 and a flexible drill string with a sectoral rotation, for example, manually using a handle and an inclination compass.

The part of the flexible drill string 4 above the ground is equipped with a sectoral rotation mechanism 15, a control inclination compass 16, a mechanism for axial feed of the device to the bottom hole (not shown in the drawing) and a flexible sleeve 17 for supplying a weak electrolyte through the channel of the tubular current conductor 7 and the tubular electrode 6 into the cavity of the electric discharge hydraulic impulse unit 5.

The flushing liquid is supplied through the gap between the walls of the pipeline 4 and current conductor 7. The liquid is delivered to the surface through the annulus.

The electric-discharge hydraulic impulse unit 5 is triggered when the electrical circuit is connected to the supply network: current conductor 7—electrode 6—weak electrolyte in a thin gap 9—housing of the hydraulic impulse unit 5—walls of the drilling string 4—neutral earthing with a feed transformer on the surface (not shown in the drawing). In this case, a powerful current discharge occurs in a thin gap 9 and forms a high-pressure vapor-ionic substance, which, increasing in volume, compresses a portion of the liquid within its physical compression ratio. When the liquid pressure reaches a critical value relative to the preset value of the valve 8 and a certain volume of high-pressure vapor-ionic substance in the cavity of the unit 5 expands, the valve opens and a jet of instantly expanded water produces a powerful ejection and blasting destruction of the bottom hole. At the same time, its side walls are loosened. Small particles of the broken mass are carried away to the surface by the flushing liquid, and large particles are crushed by subsequent shots during the driving process.

The direction of drilling is changed by turning the drilling string 4 at a certain sectoral angle using the lever 15 and the control inclination compass 16 on the surface. In this case, the drill bit 14 is deepened into the coal face by the axial feed of the drilling string 4, and the associated bit sub 13 at a sectoral rotation on the oblique dividing plane 12 will accordingly change the axial direction of the jet by a certain angle and move the electric discharge hydraulic impulse unit 5 with the drilling string 4 in the new direction.

The well is flushed through the cavity of the pipeline 4, and the flushing liquid is delivered through the annulus between the walls of the pipe string 4 and the well.

In the process of driving in coal, the new device simultaneously loosens and locally fractures the coal mass, which during operation will contribute to a more intensified gasification of high-ash coals.

In addition, powerful shock waves from pneumatic-hydraulic pulses are continuously received by the sonar receiver on the surface, and it identifies the location of the executive body in the coal seam and, at the same time, the local boundaries of the coal seam. This is used for operational control of the drift direction of both opening and gasification channels. In addition, the limited-swing flexible pipeline in the first option has the ability to freely and repeatedly change its curvature without the risk that the well will exit the coal seam.

The downhole steam generator (Fig. 6.9) is installed in the borehole collar 1 and is made of a pair of coaxially installed pipelines 2 and 3 with a sealed annulus 4, where water evaporates and steam overheats. Cold water is supplied through the central axial pipe 5, which in the lower part of the steam generator is connected to the

Fig. 6.9 Downhole steam generator schematic

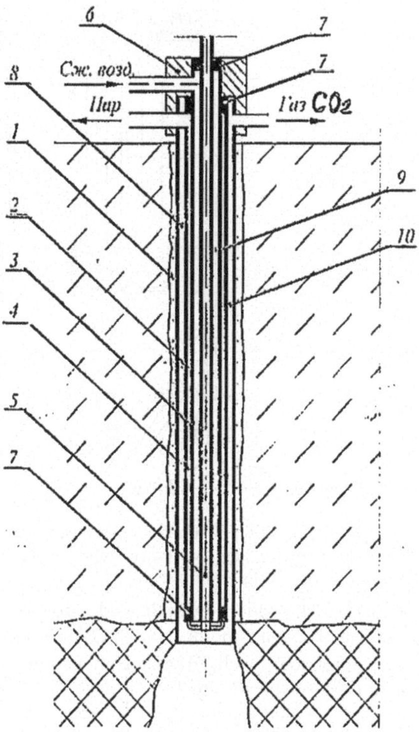

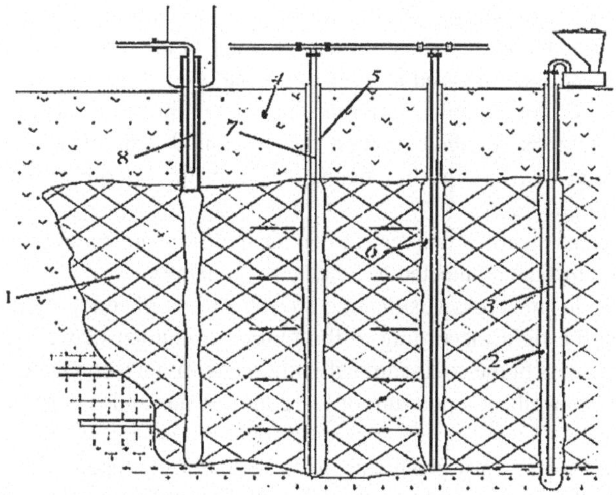

1 - Coal seam in down-dip view; 2 - Clay mud wall; 3 - Blast pipeline;
4 - Overburden; 5 - Packer; 6 - Gasification channel; 7 - Borehole collar;
8 – Underground heat-exchange steam generator

Fig. 6.10 Preparation of a gasification site by means of mutually parallel wells for filter coals. (1) Coal seam in down-dip view; (2) Clay mud wall; (3) Blast pipeline; (4) Overburden; (5) Packer; (6) Gasification channel; (7) Borehole collar; (8) Underground heat-exchange steam generator

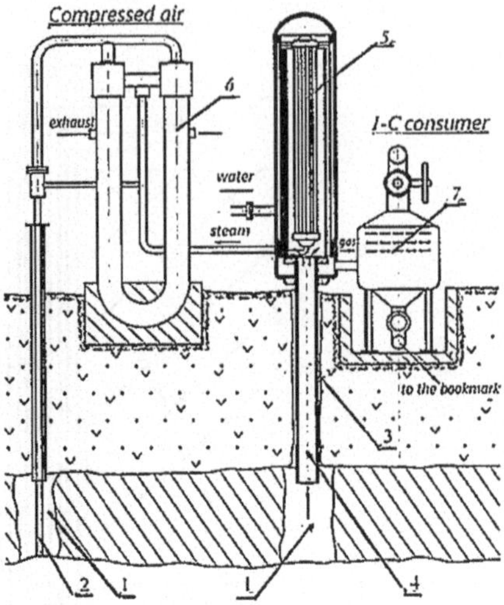

1 - Gasification channel; 2 - Blast pipeline; 3 - Borehole collar of the gas producing well; 4 - Gas combustion area; 5 - Tower-type steam generator; 6 – Hydrodynamic compressor; 7 - Cyclone

Fig. 6.11 Schematic view of the underground gasification site where coal is completely burned and superheated steam is produced. (1) Gasification channel; (2) Blast pipeline; (3) Borehole collar of the gas producing well; (4) Gas combustion area; (5) Tower-type steam generator; (6) Hydrodynamic compressor; (7) Cyclone

annulus 4. The cantilever pipe bundle is press-fitted into the head 6, and the end annuli are pressed with plugs 7. Waste combustion products are discharged through the annulus 8 to heat and evaporate formation water in the next gasification section being prepared and then released into the atmosphere.

Compressed air for complete afterburning of combustible gas at the inlet to the steam generator is supplied through the annulus 9. Complete afterburning of the gasification gas is done in the channel 10 from the bottom at the inlet to the downhole steam generator.

The superheated steam is delivered to an air hydrodynamic compressor and to a steam-hydraulic turbine connected to an electric generator.

It is possible to prepare a gasification site by means of mutually parallel wells (Fig. 6.10). On the right (according to the figure), a number of wells 2 are drilled through the thickness of the coal seam with minimal gaps between them. Clay mud is injected into these wells at a pressure of 8–12 atm. A gas and waterproofing wall is formed.

From the left side, a gas producing well is driven equipped with a downhole steam generator 8 in the zone of overburden 4. Between these boundary wells along the edge of the steeply dipping formation 1, a number of blast injection holes are drilled equipped with blast pipes 3. All wells in the overburden zone 4 are cased with steel pipes 7 with their upper ends plugged with packers 5.

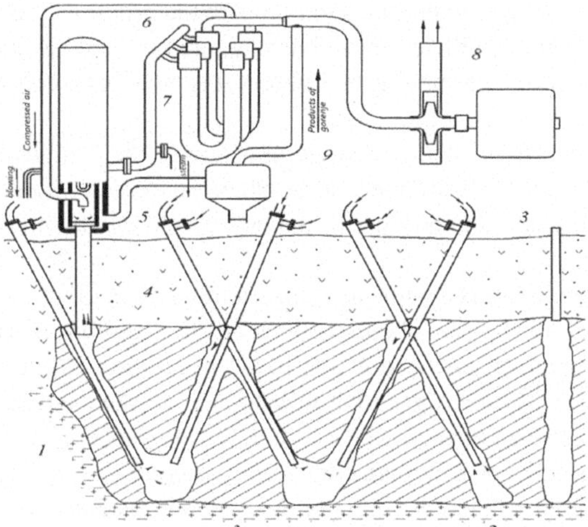

1 - Coal seam; 2 - Barrier wall made of plastic clay mud; 3 - Blast pipelines; 4 - Gasification channels; 5 - Underground tubular steam generator; 6 - Afterburner; 7 - Hydraulic percussion compressor; 8 - Gas turbine; 9 - Gas-purifying chamber.

Fig. 6.12 General arrangement of the underground gasification site where thermal energy is processed into electrical energy. (1) Coal seam; (2) Barrier wall made of plastic clay mud; (3) Blast pipelines; (4) Gasification channels; (5) Underground tubular steam generator; (6) Afterburner; (7) Hydraulic percussion compressor; (8) Gas turbine; (9) Gas-purifying chamber

Combustible gases are completely burnt before entering the downhole steam generator 8. Superheated steam from the steam generator is supplied both to service the gasification site and to generate electricity for external consumers.

Figure 6.11 shows a part of the power unit of the gasification site on its surface, which includes a steam generator 5, for example, of a tower type. The steam generator is connected to a rough cleaning device—cyclone 7. Then the ash is supplied for the backfilling of the burned-out space.

Superheated high-pressure steam is partially supplied as an energy drive to hydrodynamic compressors 6. Shocked air and exhaust steam during the return phase of the working process, are injected into the blast pipeline into the fire zone of the gasification site.

Figure 6.12 shows a schematic view of preparation for mining of a steeply dipping coal seam 1 with cross-shaped blast pipelines 4. Such an arrangement is optimal with a low filtration capacity of high-ash coals.

The site is also isolated from the rest of the formation by barrier wall 2. The gas outlet channel supplies the hot gas mixture to the downhole steam generator 5, and the combustible part of the gas mixture is completely burned in the afterburner 6 on the surface. The superheated steam is partly supplied to the hydraulic percussion compressors 7, and the other part is supplied to the steam turbine 8. The new design of the turbine 8 ensures the operation on a mixture of steam and exhaust gas, which is preliminarily re-cleaned using a gas-purifying chamber 9.

In general, the gasification products in line with the proposed technology, involving steam-air blast and complete carbon burning, are hot inert gas used in downhole steam generators to produce superheated steam. It is the working medium for mobile steam-hydraulic turbines of a new design.

The gasification process with partial combustion of the combustible part ends at the inlet to the downhole steam generators, where an excess volume of compressed air is supplied and where the burning of the entire volume of the resulting combustible gases is ensured.

Hot combustion products with an initial temperature of up to 1200 °C pass through the annuli of the downhole steam generator and are discharged into the blast pipelines of another gasification site to warm up the coal mass and evaporate the formation water. The inert gas, that was cooled, washed and purified by forced filtration through the coal mass, is released into the atmosphere. Ash and all harmful impurities remain under the ground. Since the gas comes out of the ground rather than through a tall pipe, it can be bubbled with alkali, resulting in chalk and water, thus neutralizing the greenhouse gas.

Superheated steam is supplied as a drive to hydraulic percussion compressors and steam-hydraulic turbines. If consumers are located nearby, the end product can be steam or hot water.

It should be noted that burning coal in the ground and producing electrical and/or thermal energy significantly reduces the possible change in the surface compared to its mining and delivery to the power plant, since the ash remains in the ground.

Underground Gasification Exemplified by the Coal Deposit in the East Kazakhstan Region

To obtain fuel gas of satisfactory quality, the proposed technology provides for high-pressure steam-air blasting exceeding the hydrostatic pressure at a depth of 800–850 m. The thickness of the superincumbent clay overburden is from 50 to 120 m.

The high pressure of the mixture of compressed air and steam, which is obtained by percussion adiabatic compression in special hydrodynamic compressors, ensures a high heat content of the steam-air mixture up to 2800–3000 kJ/kg.

As a result, the carbon part is intensively decomposed resulting in an ideal semi-water gas CO—40%; H_2—18%; N_2—42% (Spravochnik, 1967).

The volume of an ideal semi-water gas formed from 1 kg of carbon is 4.65 m^3. With an average ash content of coal equaling 27.5%, the volume of gas from 1 kg of coal will be 3.35 m^3. The calorific value of gas is 7.0 MJ/m^3. The amount of heat from combustion of 1 kg of coal is 23.45 MJ.

To obtain high-quality fuel gas, the proposed technology provides for steam-air blasting while using the steam as a hydrostatic drive in the process of compressing the atmospheric air by hydrodynamic compressors, and the steam is injected together with compressed air.

The blast pressure corresponds to the hydrostatic pressure of formation water along the entire depth of its development, i.e., 90–100 kg/cm^2.

Underground gasification parameters are shown in below table.

No.	Parameter	Unit	Value
1	Total heat consumption of the northern and central regions	MJ	69×10^6
2	Gas output from underground gasification per 1 kg of coal	m^3/kg	4.63
3	Estimated calorific value of dry gas, including per 1 m^3 of gas, Per 1 kg of coal	kJ/m^3 kJ/kg	5650 26,160
4	Equivalent gas production for the region per year	Bln m^3 m^3/min	12.285 23,250
5	Annual volume of mining at the coal deposit	Mln. T	3.564
6	Blast air consumption, including per 1 kg of coal; In 1 min	m^3/kg m^3/min	2.0 10,030

Table shows that in order to provide a city of about 200,000 people with thermal energy, it is necessary to burn 24,500 tons of coal a year using the borehole method.

The purified inert greenhouse gas can be bubbled with alkali at the exit from the well to produce chalk and water, thereby achieving carbon neutrality.

6.8 Conclusion

The problem of wind and sun energy variability can be solved by combining two methods: the accumulation of energy and the increase in its production without the rise or insignificant rise in its cost.

Thus, the energy production by wind farms is increased due to a significant increase in time utilization factors and the use of rated power while maintaining a high efficiency factor without the increase or even with the reduction of manufacturing costs. Combined wind and solar installations, which increase energy production, are also getting ground. But they are used for consumer-grade sources. These involve lithium-ion batteries, which have a short service life, are expensive and based on the use of very rare metals. Therefore, there appear batteries that work on widely used elements, such as iron-air, zinc-bromine, etc.

In this regard, pumped storage units are being developed and even already being used for industrial scales. Gravity accumulators that can maintain the continuity of energy supply from industrial wind and solar installations, have a long service life and can be easily disposed of are becoming promising.

The deficit of very rare and expensive metals is also solved by their multiple reduction without loss in the efficiency of renewable energy sources. Production of solar panels ensures not only a 1000-fold reduction in elements, but also the use of more efficient materials, such as silicon carbide (SiC) and gallium nitride (GaN), the efficiency of which is three times higher than that of widely used silicon (Farnell, 2022), etc.

The use of the proposed designs and the above-mentioned materials in wind energy, a 1000-fold reduction and replacement of photovoltaic cells, practically eliminate the problem of recycling of used wind farms and solar panels.

In addition, the energy consumption at the manufacturing of the proposed wind farms and solar panels not only does not exceed the amount of energy they generate, but also pays off many times over when they are used. And the opinion that the cost of energy generated by renewable sources is higher compared to traditional sources is already becoming obsolete.

Regarding to coal energy, it is shown that there is a real possibility for the effective industrial development of steeply falling coal deposits. Only in Kazakhstan such explored and forecasted reserves of hard coal make up to 7.6 billion tons. Its can be involved in mining by the above-mentioned technology with power generation, which excludes environmental contamination and increases decarbonization.

Thus, it is possible to improve thermal power plants, which practically do not harm the environment, to reduce the cost of electricity by avoiding the deposit development, coal transportation to the station and use of significant areas of farmland for waste rock dumps and ash landfills.

In general, a gradual transition to renewable energy sources and the in-situ use of mineral fuels are practically the main ways to maintain the ecological balance on Earth.

Acknowledgments The chapter was written based on the results of the research funded by the program-oriented and grant financing by the Ministry of Education and Science of the Republic of Kazakhstan (No. BR05236263, No. APO5131126).

References

Bakulev, G. D. (1955). *Razvitie ugolnoi promishlennosti Doneckogo basseina* (p. 502). M. GIRL.

Bikbaev, R. G., Vetrov, S. Y., Timofeev, I. V., & Shabanov, V. F. (2022). Nanoparticle shape optimization for Tamm-plasmon-polariton-based organic solar cells in the visible spectral range. *Photonics, 9*(11), 786. https://doi.org/10.3390/photonics9110786

Bradley, C.S. (1885). Patent USA №312.802.

Buktukov, N.S., Buktukov, B.J., Moldabaeva, G.J., & Zhakip, A.K. (2011). Generator (varianti) patent RK № 26968.

Buktukov, N.S., Buktukov, B.J., Moldabaeva, G.J., & Zhakip, A.K. (2019). Vetroelektrostanciya. Patent RK №31791.

Buktukov, N.S., & Gumennikov, E.S. (2022). Alternativnaya energetika v reshenii problem dostijeniya uglerodnoi neitralnosti (obzor na primere kazahstanskih uslovii) Sbornik tezisov dokladov "Innovacii v alternativnoi energetike: proizvodstvo i nakoplenie Vserossiiskaya konferenciya s mejdunarodnim uchastiem" (Dubna, 14–16 noyabrya 2022 g).

Buktukov, N.S., Kusainov, A.S., & Kusainov, S.G. (2016). Solnechnaya fotoelektricheskaya batareya (varianti,) Patent RK № 31796.

Farnell. (2022). Wide-bandgap semiconductor: The future of SiC and GaN technology.

Geologiya mestorojdenii uglya i goryuchih slancev SSSR. (1973). Kn.1-Moskva: Nedra, 1973 g. T. 5.

Kiptilii V.G., & Gusinskii G.M. (1987). Ustroistvo dlya preobrazovaniya energii vidimogo i infrakrasnogo izluchenii v elektricheskii tok. Patent ER M 215482., kl. H.01 L 27/14, 1987.

Mendeleev, D. I. (n.d.). Buduschaya sila, pokoyaschayasya na beregah Donca. *Soch, 19*, 568–569.

Saravanapavanantham, M., Mwaura, J., & Bulović, V. (2022). Printed organic photovoltaic modules on transferable ultra-thin substrates as additive power sources. *Small Methods, 7*(1), 2200940. https://doi.org/10.1002/smtd.202200940

Spravochnik himika. (1967). Leningr. Otd.: Himiya, T. 6.

Vaidya, N., & Solgaard, O. (2022). Immersion graded index optics: Theory, design, and prototypes. *Microsystems & Nanoengineering, 8*, 69. https://doi.org/10.1038/s41378-022-00377-z

Chapter 7
How Natural Resources, Corruption, Globalization, and CO_2 Emissions Affect the Economic Growth in Europe?

Victor Moutinho, Henrique Espinosa Oliveira, Helena Santos Oliveira, and Filipa Vasco Vaz

Abstract This study investigates the impact of economic, political, social, and environmental variables on economic growth in the Member States of the European Union. A study was carried out based on scoreboard data for the 27 Member States of the European Union from 1995 to 2019, using Panel-Corrected Standard Errors in the first phase and DOLS (Dynamic Ordinary Least Squares), FMOLS (Fully Modified Ordinary Least Squares), PMG (Pooled Mean Group), and DFE (Dynamic Fixed Effect) estimators in the second phase. The static analysis results show that for all coefficients associated with the selected drivers for economic growth, significant statistical evidence is observed. Moreover, according to the results in the second phase, the long-term results of the estimators show that the variables CO_2 emissions and the corruption perception index positively influence the economic growth of different countries. In contrast, the variables Total Incomes of Natural Resources and Jure Political Globalization Index negatively influence it. In the short term, in the DFE estimator, the only variable that negatively influences the economic growth of different countries is Total Natural Resource Rents and referring to the FMOLS and DOLS estimators, the only variable that negatively influences economic growth is the Jure Political Globalization Index.

V. Moutinho (✉)
NECE – Research Center for Business Sciences and Management and Economics Department, University of Beira Interior, Covilhã, Portugal
e-mail: ferreira.moutinho@ubi.pt

H. E. Oliveira · F. V. Vaz
NECE – Research Center for Business Sciences and Management and Economics Department, University of Beira Interior, Covilhã, Portugal
e-mail: henrique.oliveira@ubi.pt

H. S. Oliveira
CEOS-PP, Polytechnic Porto, School of Accounting and Administration, NECE – Research Center for Business Sciences and Management and Economics, Covilhã, Portugal
e-mail: oliveira@iscap.ipp.pt

© The Author(s), under exclusive license to Springer Nature Switzerland AG 2024
T. Devezas et al. (eds.), *Global Energy Transition and Sustainable Development Challenges, Vol. 1*, World-Systems Evolution and Global Futures,
https://doi.org/10.1007/978-3-031-67583-6_7

7.1 Introduction

The concept of economic growth is frequently limited to a quantitative expansion of productive capacity, often overlooking the transformation in the qualitative aspects of the economic structure. The economic growth theory aims to identify the key drivers that significantly influence this phenomenon.

The European Union's environmental policies aim to improve resource efficiency, so by 2050 Europe has three main objectives to achieve, which are to conserve, protect and strengthen the European Union's natural capital, make it a green, competitive and low-carbon economy that is, resource-efficient, and ultimately protect EU citizens from environmental pressures and reduce risks to their health and well-being. Natural resources must be used with other factors of production to generate green energy that brings economic and environmental efficiency to all countries.

Globalization, typically denoting the integration of economic activities worldwide via the exchange of goods, services, capital, and human resources, enhances the efficient distribution of resources globally and facilitates the dissemination of technological advancements. Corruption, encompassing any unethical practices or behavior, constitutes a pivotal element in the context of the relationship between "globalization and development." This relationship, being a comprehensive concept, encompasses not only economic growth but also other facets relevant to a nation's prosperity, including education and health levels.

The conventional perspective posits it is widely recognized that corruption has a detrimental effect on economic development. However, there are contrasting studies that propose corruption could stimulate development. For instance, corruption might ease business transactions in highly regulated environments. Additionally, the relationship between globalization and corruption remains ambiguous—it is uncertain whether globalization mitigates corruption by fostering transparency or if globalization, conversely, propagates corrupt practices.

Globalization can be characterized as the amalgamation of distinct nations in terms of economic, social, and political interactions. As integration among countries has progressed, there has been a growing need for quantifying and evaluating globalization. The KOF Globalization Index, initially introduced by Dreher (2006) and further refined by Gygli et al. (2019), has emerged as a prominent metric within scholarly discourse. Notably, Gygli et al. (2019) have made significant advancements in the analysis of globalization by revisiting the original KOF Globalization Index. The revised KOF Globalization Index encompasses three dimensions, delineating the assessment of globalization into economic, political, and social domains. Additionally, they have introduced de jure and de facto measures of globalization, representing a pivotal development. These novel measures have substantially improved the analysis of the globalization phenomenon and notably mitigated biases in the outcomes (Martens et al., 2015). Consequently, the three dimensions can be succinctly summarized as follows: (1) The economic dimension encompasses variables related to the flows of goods, capital, services, and market exchanges; (2) The social dimension comprises variables reflecting the circulation of ideas, information, and people; and (3) the political dimension includes variables depicting the dissemination of government policies and measures (Leal et al., 2020). The two new measures of globalization can be defined as follows: de jure globalization denotes a measure of policies that facilitate and regulate flows and activities, while de facto globalization quantifies actual flows and activities (Leal et al., 2020).

Regarding the correlation between "globalization and corruption," the prevailing viewpoint suggests that globalization diminishes corruption by introducing accountability mechanisms enforced by international trade partners. Consequently, countries actively engaged in numerous international institutions tend to exhibit lower levels of corruption. Many international bodies regard globalization as a potent tool in combating corruption, assuming it necessitates structural and institutional reforms like market liberalization, competition enhancement, technology transfer, protection of property rights, upholding the rule of law, ensuring transparency, and enhancing the quality of public services. These reforms serve as the conduit through which globalization impacts corruption.

However, there exists a body of literature positing that globalization amplifies corruption, particularly in low per capita income countries, as incoming foreign direct investment (FDI) flows may be linked to corruption, likely due to relatively higher FDI rents in these nations. The increased interactions at the international level can pose challenges, as corruption tends to be contagious. Hence, in the context of globalization, countries of origin may inadvertently export corrupt practices.

The abundance of natural resources, including oil, coal, and natural gas, along with political and economic stability, has attracted significant foreign investments into the primary industrial sectors of the economy. This influx has notably contributed to the country's economic green growth. However, it is crucial to emphasize the need to preserve these valuable natural resources.

Moreover, examining the relationship between GDP (Gross Domestic Product) per capita and the corruption perception index reveals that as a society becomes wealthier, it becomes less likely for governments to misuse their public power for private gains. The overall affluence of society plays a significant role in combating corruption. Therefore, economic growth is instrumental in reducing corruption, and lower levels of corruption, in turn, can lead to an increase in the country's GDP per capita.

In recent years, a notable trend has emerged where global CO_2 emissions have stabilized, even as the Gross Domestic Product (GDP) continues to grow on a global scale. The International Environmental Agency (IEA) attributes this levelling off of emissions to a decrease in CO_2 emissions specifically related to electricity generation within advanced economies. This decline can be attributed to the increased uptake of renewable energy sources like wind and solar power, the transition from coal to natural gas, and the increase in atomic energy production. Consequently, CO_2 emissions in advanced economies have decreased to levels akin to those witnessed in the late 1980s. Notably, this pattern extends to rapidly advancing economies like India and China, where emissions per capita have been controlled despite robust economic growth.

However, achieving relative decoupling alone is inadequate, particularly in a global context where a significant reduction in CO_2 emissions is imperative to effectively mitigate global warming and align with the objectives specified in the Paris Agreement, an international treaty to constrain global warming to less than 2 ° C. It becomes evident that significant reductions in greenhouse gas emissions are imperative to effectively address climate change.

The European Union already has well-defined targets for the percentage of total energy consumption that must be renewable. This target was 20% for 2020 and 32% for 2030. The European Union (EU) has outlined two primary objectives for its environmental initiative until 2030. These goals include a reduction of greenhouse gas emissions by a minimum of 55% in comparison to 1990 levels, accompanied by a simultaneous enhancement of energy efficiency by no less than 32.5%. These intermediate objectives are directly linked to the main objective of the European Union, which aims to make the European continent climate-neutral by 2050 (European Commission, n.d.). However, questions arise about the impact this transition could have on total economic output in the region. If energy costs become significantly higher, all economic activities dependent on the use of energy will become more expensive, which will probably reduce the population's standard of living and make social mobility more difficult.

This work contributes to the literature by offering an in-depth analysis of the relationship between economic growth and several of its key determinants: CO_2 emissions, corruption, natural resources, and globalization for European countries. Our new economic growth model incorporates the impacts of KOF globalization across economic, social, and political dimensions. The primary objective is to elucidate the trade-off that exists between economic growth and these globalization metrics. Consequently, this study holds importance for devising well-informed strategies regarding the integration of globalization across various dimensions, particularly in achieving economic growth goals within integrated European nations. Additionally, our proposed relationship underscores the potential for significant interactions between globalization's diverse dimensions and economic growth. These interactions are pivotal for advancing the growth of these economies and are central in formulating policies and strategies that support both economic growth and the influence of globalization.

It aims to: (1) analyze the consequences of CO_2 emissions on the development of the economy; (2) the study scrutinize the correlation between economic growth and the Corruption Perception Index (CPI); (3) examine the link between natural resources and financial development; (4) evaluate how de jure unbundled globalization contributes to economic growth. The factors affecting economic growth for various groups of countries have already been examined using a wide variety of methodologies for empirical research. For example, Pejović et al. (2021) utilized data encompassing 27 European Union countries and the Western Balkans over the period from 2008 to 2018. Their analysis revealed that CO_2 emissions-induced environmental degradation has detrimental effects, notably exerting a substantial negative influence on the economic growth of the examined countries. In the study conducted by Uberti (2021) covering the period from 1790 to 2010 and focusing on

European countries, researchers investigated the impacts of corruption on economic growth. The findings provided substantial evidence indicating that corruption has a detrimental effect on economic growth. On the other hand, authors such as Anyanwu et al. (2021) and Rahim et al. (2021) tried to understand whether abundant natural resources increase the negative impact of income inequality and, consequently economic growth. Both of these studies concluded that countries with more natural resources tend to be less equal and that the incomes of these resources tend to harm GDP, thus inhibiting the economic growth of the nations of these countries. (Swadźba, 2020) concluded that globalization harms economic growth and socio-economic development in a range of countries, while Ayomitunde et al. (2020) concluded that the metric variables of globalization stimulate economic growth in European countries. According to the study by Radulović and Kostić (2020), the short-term results of the study suggest a distinction in the effects of globalization dimensions on economic growth. Economic globalization and social globalization were found to have a positive impact on economic growth, while political globalization exhibited a negative impact. In the long term, existing evidence supports the notion that economic globalization positively impacts economic growth. Conversely, social globalization and political globalization have been shown to exert a negative influence on economic growth.

The structure of the chapter is outlined as follows: Section 7.2 offers a succinct review of the literature, Sect. 7.3 delineates the data and methodologies employed, Sect. 7.4 presents the findings and ensuing discussions, and Sect. 7.5 furnishes concluding remarks encapsulating the principal findings.

7.2 Literature Review

7.2.1 Economic Growth: Corruption

Although it is almost intuitive in the literature that the effect of corruption negatively affects a country's economic growth, many authors have studied this relationship and tried to find solutions. Corruption is generally viewed as a detrimental factor influencing economic development (Aidt, 2009; Amorim Janielly, 2018; d'Agostino et al., 2016; Kaufmann, 1997; Khumawala & Ramchand, 2005; Mauro, 1995; Murphy et al., 1991). However, some studies present a contrasting perspective, suggesting that corruption can stimulate economic growth, aligning with the "Greasing the Wheels" hypothesis (Djankov et al., 2002; Dreher & Gassebner, 2013; Huntington, 1968; Leff, 1964).

The author (Mauro, 2004) tried to understand why corruption persists over time, despite its negative impact on countries' economic growth. Using various models to explain this relationship, he concludes that this phenomenon becomes generalized and individuals have no incentive to put an end to it, although societies would be better off without the presence of corruption.

Li and Wu (2010) carried out a study for 53 countries in two different time periods and the authors (Moiseev et al., 2020) used panel data for 45 countries, where they claim that confidence in a country's corruption (efficiency ratio) mitigates its negative effect on economic growth, since they wonder why some countries achieve high levels of growth in the presence of corruption. Thus, the more wealth a society has, the less corruption is present and the higher the economy's GDP per capita.

The study carried out by Uberti (2021), based on the period 1790–2010 for European countries, aims to insert a V-Dem indicator to examine the effects of corruption on economic growth, which will fill in the long-term gaps in Corruption Perception Index data. The study concludes that corruption negatively affects a country's economic growth.

7.2.2 Economic Growth: Natural Resources

The extraction of natural resources is increasing all over the world, so it is necessary to study the relationship between countries' economic growth and natural resource rents to see if they have a positive or negative impact on GDP per capita. The authors (González-Val & Pueyo, 2019) state that the consequence of increased exploitation of countries' natural resources has been growth in the world's population. Using a model for just two countries, they concluded that revenue from the extraction of natural resources has increased greatly over the years, enabling countries to increase their GDP value.

On the other hand, authors such as Anyanwu et al. (2021) with panel data for the years 1988–2012, and Rahim et al. (2021) with data from 11 countries for the years 1990–2019, try to understand whether abundant natural resources increase the negative impact of income inequality on the economic growth of countries. Both conclude that countries with more natural resources tend to be more unequal and that the rents from these resources tend to have a negative impact on their GDP, thus inhibiting the economic growth of the nations in these countries.

Amini (2018) used 22 developed countries and 61 developing countries for his research for the period between 1996 and 2010. This study concludes that the existence of large quantities of natural resources does not necessarily mean economic and political development, since the abundance of resources can have advantages or even risks for a nation. Although in recent decades most studies have shown that natural resources serve to increase economic growth (Van Der Ploeg, 2011) and (Frankel, 2010), more studies should be carried out with various groups from different countries to understand the relationship between natural resources and economic growth.

7.2.3 Economic Growth: Globalization

Globalization can be defined as a process of comprehensive economic integration on a global scale, erasing traditional economic boundaries. Key indicators often utilized to quantify globalization include the KOF Globalization Index, foreign direct investment (FDI), and international trade. Nonetheless, achieving a consensus regarding the impact of these variables on the economy remains elusive. The KOF Globalization Index is acknowledged and cited in academic literature, as evidenced by studies such as those conducted by Bilgili and Ulucak (2020); Danish et al. (2020); Gozgor et al. (2020); Guan et al. (2020); Leal et al. (2020); Liu et al. (2020); Sethi et al. (2020); and Woo and Jun (2020). The KOF globalization index can be divided between de facto and de jure measures along the different dimensions of globalization, which are economic, social, and political. Gygli et al. (2019) introduced the revised version of KOF, across economic, social, and political aspects to assess the effect of globalization metrics on economic growth. Noted as significant the results support that de facto and de jure globalization present differentials of impact on economic growth, given that increases in economic growth are induced by the effect of the increase in globalization effect, according to Swadźba (2020) the metrics of globalization the economic and political level de jure and social globalization have greater impacts on economic growth.

More specifically, Huisman and Kiliç (2015) analyze the relationship between globalization and economic growth using the KOF index, distinguishing between economic, political, and social globalization. Some studies focus on 71 countries from 1981 to 2011. Their results indicate that economic and political globalization has improved economic growth in some developing countries, while social globalization has undermined economic growth in some. Olimpia and Stela (2017) found similar results in their work carried out in Romania from 1990 to 2013.

The association between globalization and economic growth is also analyzed by Majidi (2017) using the KOF index with a distinction between economic, political, and social globalization. A study was carried out on 100 countries from 1970 to 2014. Majidi (2017) found negative effects on economic growth for political globalization in upper-middle-income countries, while the effects of economic and social globalization were not significant. Focusing on lower-middle-income developing countries, Majidi (2017) also found that political globalization increases economic growth, but economic and social globalization do not affect economic growth in these countries. Using the generalized method of moments (GMM), Samimi and Jenatabadi (2014) examined the association between economic globalization and economic growth in Organization for Economic Co-operation countries. Their findings revealed that economic globalization, as quantified by the KOF index, fosters economic growth in high- and middle-income countries but has a detrimental impact on low-income nations. In a study conducted by Radulović and Kostić (2020) involving 19 eurozone countries, it was revealed that economic and social globalization had a positive short-term impact on economic growth, while political globalization exhibited a negative impact. However, in the long run, only economic globalization displayed a positive effect on the economy. Another study by Gasimli et al. (2022) demonstrated a positive relationship between KOF Globalization metrics and sustainable growth in the CIS region. Additionally, Haini et al. (2023) explored the relationship between the Social KOF index, tourism, and economic growth across 143 countries. They discovered that the Social KOF index positively influenced tourism, subsequently contributing to economic growth.

Moreover, Santiago et al. (2020) investigated the impact of KOF globalization indicators, economic freedom, and energy consumption on economic growth in Latin America and the Caribbean. Their study uncovered the positive effects of globalization, particularly its social and economic indicators, on economic growth. However, the political indicator was associated with a negative impact. Furthermore, economic freedom was correlated with a decrease in economic growth, while energy consumption showed a positive association with economic activity.

Certain scholars investigate the correlation between globalization and economic growth through specific indicators like foreign direct investment (FDI), trade openness, and official development assistance. Using the dynamic growth model, Zahonogo (2018) analyzed the relationship between trade openness (economic globalization) and economic growth in sub-Saharan Africa. Based on panel data, Zahonogo (2018) demonstrated a nonlinear relationship between globalization and economic growth and recommended that SSA (sub-Saharan African) countries control trade openness, especially imports of consumer goods, to stimulate their economic growth. Egbetunde and Akinlo (2015) examined the effects of financial globalization on the economic growth of sub-Saharan African countries. Using an error-correction panel model, the authors concluded that SSA economies will benefit from financial globalization in the long term if governments promote sound macroeconomic policies and strong institutions. Musila and Yiheyis (2015) distinguished between general trade openness (de facto globalization) and trade policy-induced openness (de jure globalization) to analyze the effects of globalization on growth in Kenya. They concluded that de facto globalization does not affect growth, while de jure globalization negatively affects economic growth in Kenya (Hodabalo Bataka, 2019). This study complements the existing literature by adopting a methodological approach to solve the problem of cross-sectional dependency and by making a perfect distinction between de facto and de jure globalization in the context of sub-Saharan African countries.

Moghaddam and Redzuan (2012) used FDI to analyze the effects of globalization on the economic development of developing countries. They concluded that de facto globalization does not affect growth, while de jure globalization negatively affects economic growth in Kenya. Their study led to the conclusion that attracting FDI supports economic growth. Borensztein et al. (2005) obtained similar results in their work on 69 developing countries. Khaliq and Noy (2007), on the other hand, concluded that aggregate FDI has a positive impact on economic growth in Indonesia, while sectoral analysis reveals the negative effects of FDI directed towards the mining sector.

Ayomitunde et al. (2020) used as a study base data from European countries from 1990 to 2018 and concluded that the variables of the globalization index (net inflows of FDI—foreign direct investment and trade opening), brought economic growth to European countries. This shows that the impact of globalization has been positive.

7.2.4 Economic Growth: Greenhouse Gas Emissions

Economic growth has traditionally been associated with a rise in carbon dioxide (CO_2) emissions and energy consumption, giving rise to the perception that a more prosperous world comes at the expense of the natural environment and climate (González-Álvarez & Montañés, 2023). Academic discussions have been vibrant regarding the potential for achieving complete decoupling between emissions and growth. This involves the notion that, with the adoption of clean energy technologies, emissions can decrease even as economic growth remains uninterrupted (González-Álvarez & Montañés, 2023). The most developed parts of the world have begun to harmonize their patterns of economic growth and energy consumption with their environmental policies (González-Álvarez & Montañés, 2023).

In a study by Gyamfi et al. (2020), a positive relationship was noted between economic growth, nonrenewable energy consumption, and CO_2 emissions within E7 economies.

In studies by Shahbaz et al. (2015) and Shahbaz et al. (2016) focusing on India, the authors investigated the interconnections between globalization, energy consumption, economic growth, and environmental degradation. These studies supported the Environmental Kuznets Curve (EKC) hypothesis and identified causal relationships akin to those observed in the study by Haseeb et al. (2018). Additionally, in the study by Shahbaz et al. (2016), an endogenous approach was employed to account for economic growth, financial development, and urbanization from 1971 to 2012.

Regarding the Chinese economy, Shahbaz et al. (2017) conducted a study examining the relationships between CO_2 emissions, economic growth, coal consumption, and globalization. They employed cointegration and Vector Error Correlation Model (VECM) approaches in their analysis. The outcomes substantiated the Environmental Kuznets Curve hypothesis, indicating a relationship characterized by there exists an inverted U-shaped relationship between economic growth and environmental degradation. Additionally, the study found that coal consumption exacerbated environmental degradation, while globalization was associated with an improvement in environmental quality for the examined sample.

One notable study by Shahbaz et al. (2018) investigated the influence of economic growth and related factors on environmental degradation over the period from 1970 to 2014. The findings of this study suggested that all the examined variables had detrimental effects on the environment.

However, in a study focused on Italy, conducted by Akadiri et al. (2019), the relationship between economic growth and environmental degradation was reported to be insignificant. In essence, the results from this study indicated that, at least in the case of Italy, economic growth did not have a harmful impact on the environment.

In a study conducted by Baek et al. (2009), the research analyzed the impact of trade liberalization and economic growth on environmental degradation across 50 developed and developing economies from 1960 to 2000. The study's outcomes supported the EKC hypothesis. Specifically, the findings demonstrated a positive effect of trade liberalization on environmental quality in developed economies, while a negative effect was observed in developing economies. These results align with the hypothesis proposing the existence of pollution havens in developing economies.

In another study by Rahman (2020), the research focused on investigating the effects of electricity consumption, economic growth, and globalization on carbon dioxide emissions within the top 10 largest electricity-consuming economies. The study provided evidence affirming the presence of the EKC when analyzing the entire panel of countries. However, the EKC curve was specifically identified for Japan, Germany, and South Korea when each economy was examined individually. Notably, the study revealed that electricity consumption had an adverse impact on the environment, except in China, India, Japan, the United States, and the United Kingdom. Furthermore, the findings indicated a negative influence of globalization on CO_2 emissions across the entire panel, particularly notable for China, the United States, and the United Kingdom when assessed on an individual basis.

Between 1970 and 2015, Liu et al. (2020) investigated the relationship between globalization and CO_2 emissions for the G7 economies, representing the seven most industrialized economies globally, including Germany, Canada, the United States, France, Italy, Japan, and the United Kingdom. The study provided evidence supporting the Environmental Kuznets Curve hypothesis, suggesting that an increase in GDP leads to a subsequent increase in emissions. Additionally, an augmentation in the consumption of renewable energies tends to correlate with a reduction in emissions. Similarly, Wang et al. (2020) validated the EKC hypothesis for the G7 economies by exploring the relationship between energy production through biomass, environmental degradation, globalization, and economic growth. Their findings revealed that an increase in energy production through biomass results in a reduction in environmental degradation, consistent with Liu et al. (2020) results regarding the relationship between renewable energy consumption and emissions. Furthermore, the causality test conducted by Wang et al. (2020) demonstrated a one-way relationship between biomass production and economic growth, and globalization to environmental degradation.

For the American economy, Usman et al. (2020) concluded that in the long run, economic growth and the consumption of renewable energies mitigate environmental degradation, while globalization and financial development contribute to environmental destruction. However, in the short term, only the consumption of renewable energy reduces environmental degradation. The findings from Pata and Yilanci (2020) research suggest that globalization and the adoption of renewable energy sources are crucial factors in mitigating environmental degradation within the United States economy. Moreover, the research concludes that the Environmental Kuznets Curve hypothesis holds for the United States from 1980 to 2016.

The study by Mert and Bölük (2016) investigated the relationship between greenhouse gas emissions, foreign direct investment (FDI), economic growth, and energy consumption for a sample of 21 countries adhering to the Kyoto Protocol. Utilizing the Pooled Mean Group and Mean Group methods and causality testing, the authors found that the Environmental Kuznets Curve hypothesis was not supported in this sample. Causality tests indicated a two-way relationship between economic growth and FDI, economic growth and greenhouse gas emissions, economic growth and fossil energy consumption, emissions and fossil energy consumption, and a unidirectional relationship from fossil energy consumption to renewable energies, as well as from emissions to renewable energy consumption.

Sarkodie and Strezov (2019) investigated the relationship between economic growth, FDI, energy consumption, and greenhouse gas emissions for developing economies, specifically China, India, Iran, Indonesia, and South Africa, to test the hypotheses of pollution havens and EKC. Their results validated both hypotheses and suggested that FDI promotes economic growth. Udemba et al. (2020) studied the relationship between FDI, carbon emissions through energy consumption, and China, finding evidence to support the pollution haven hypothesis. Causality tests showed a one-way relationship between foreign investment, energy consumption, entry of tourists, and economic growth. Some indications of increased industrial activity caused by FDI led to higher energy consumption. In a study by Peng et al. (2016), the causality between CO_2 emissions, FDI, and economic growth in China's provinces from 1985 to 2012 was examined. Their results indicated a two-way relationship between FDI and economic growth in only one province, while unidirectional causality from FDI to GDP was observed in five provinces and reverse causality in only one. The causality test also showed that economic growth increased emissions in five provinces, and a two-way relationship was observed in only one province. Similar results were found between FDI and CO_2 emissions.

7.3 Materials and Methods

7.3.1 Panel-Corrected Standard Errors (PCSE)

Estimating panel data proves advantageous in mitigating collinearity among explanatory variables, providing increased degrees of freedom and enhancing estimation efficiency. Panel data facilitates the inclusion of effects not solely accounted for in cross-sectional or temporal data, a critical aspect in addressing econometric challenges like unobserved individual heterogeneity (Wooldridge, 2002). This pertains to unrecorded variables that correlate with the explanatory variables, potentially resulting in inconsistent estimates.

When the unobserved element is unrelated to the explanatory variables, it is termed an "individual random effect," leading to the adoption of the Random Effects Model (REM) or the pooled OLS model. The choice between these models depends on the variability of the unobserved effect, which can be evaluated using tests like the Lagrange Multiplier Test. Conversely, when the unobserved element correlates with the explanatory variables, it is denoted a "fixed effect," warranting the use of the Fixed Effects Model (FEM) (Wooldridge, 2002).

In this study's first empirical stage, various specifications for the Panel-Corrected Standard Errors (PCSE) method were considered, including CORR (IND), AR (1), PSAR (1), and HET-AR (1). CORR (IND) signifies a contemporaneous correlation PCSE model, while AR (1) involves the PCSE estimator considering first-order serial correlation. PSAR (1) extends AR (1) by incorporating a panel-specific AR process coefficient (1), assuming diverse coefficients for each panel. On the other hand, HET-AR (1) assumes heteroscedasticity and common coefficients for the first-order autocorrelation process across panels.

In the second stage of the empirical analysis, four distinct estimation methods were employed to ascertain the proposed cointegration relationship. This approach was aimed at comparing outcomes across these methods, thereby enhancing the robustness and reliability of the estimates (Wooldridge, 2002).

7.3.2 Unit Root Test

The empirical analysis begins with subjecting the variables to panel unit root tests to ascertain their stationarity status. Unit root tests are commonly categorized into two major groups: first-generation tests that assume cross-sectional independence, and second-generation tests that explicitly consider specific forms of cross-sectional dependence (Pesaran, 2007).

To ascertain, within the framework of the null hypothesis, whether all individual panel series exhibit a unit root, we utilized the test introduced by Pesaran (2007), applying the following panel-based Augmented Dickey-Fuller (ADF) test. This test imposes parameter constraints by maintaining their uniformity across the sectional regions (refer to Eq. (7.1)):

$$\Delta y_{it} = c_i + \rho_i \cdot y_{it-1} + \sum_{j=1}^{k} c_j + \rho_i \cdot y_{it-j} + \varepsilon_{it} \tag{7.1}$$

The test introduced by Pesaran (2007) offers a simplified approach to testing, wherein the standard Augmented Dickey-Fuller (ADF) estimations are expanded by incorporating lagged levels and first differences for each series in the common ADF estimations (Pesaran, 2007). The fundamental panel stationarity test is based on the average of these cross-sectional augmented statistics, termed CADF (Pesaran, 2007). This statistic can be utilized to derive various versions of the *t*-bar, the *P* test, and the inverse normal test. The test, known as CIPS, is straightforward and conceptually clear (Pesaran, 2007). CIPS stands for Cross-sectionally Augmented IPS. Notably, due to specific characteristics, such as the null distribution of each CADF, these statistics are asymptotically independent nuisance bounds (Pesaran, 2007). The cross-sectionally augmented adaptation of IPS is expressed as follows.

$$y_{i,t} = (1 - \delta_i)\mu_i + \delta_i y_{i,t-1} + v_{i,t} \tag{7.2}$$

In which, $v_{i,\ t} = y_i f_t + e_{i,\ t}$, f_t is the observed common factor and $e_{i,\ t}$ is the individual specific-error. To calculate the unit root we can reuse the equation, $\Delta y_{i,\ t} = \alpha_i \beta_i y_{i,\ t\ -\ 1} + \gamma_i f_t + e_{i,\ t}$, where $\alpha_i = (1 - \delta_i)\mu_i$, $\beta_i = -(1 - \delta_i)$ and $\Delta y_{i,\ t} = y_{i,\ t} - y_{i,\ t\ -\ 1}$. That said, we can present the test hypotheses, $H_0 : \beta_i = 0$ for all i, versus $H_1 : \beta_i < 0$ for $i = 1, 2, \ldots, N_1$ $\beta_i = 0$ for $i = N_1 + 1, N_1 + 2, \ldots N$ (Altıntaş & Kassouri, 2020). The CIPS statistic is given by

$$CIPS = \frac{1}{N} \sum_{'=1}^{N} CADF_i \tag{7.3}$$

The $CADF_i$ statistic will be not cross-sectionally independent, overdue the presence of a common factor. This test has reliable settings and can report for serial correlation (Altıntaş & Kassouri, 2020).

7.3.3 Cointegration Test

Pedroni (1999) elaborates on the asymptotic and finite-sample properties of the test statistics, aiming to investigate the null hypothesis of non-cointegration within the panel. Consider the following model:

$$y_{it} = \alpha_i + \delta_{it} + X_{it}\beta_i + e_{it} \tag{7.4}$$

The parameters α_i and δ_i account for potential member-specific fixed effects and deterministic trends, respectively. Moreover, the slope coefficients β_i are flexible, allowing for potential variation across individuals. Consequently, the cointegrating vectors can exhibit heterogeneity across panel members, encompassing both the long-run cointegrating vectors and the dynamics. This approach acknowledges the likelihood of differing parameters across countries (Lee et al., 2008).

Pedroni (1999) proposed tests can be categorized into two groups. The first category employs an approach within the dimension of the study incorporating four statistics: the panel v-statistic, the panel ρ-statistic, the panel PP-statistic, and the panel ADF-statistic (Lee et al., 2008). These statistics combine the autoregressive coefficients across diverse members to conduct unit root tests on the estimated residuals (Lee et al., 2008).

The second category, proposed by Pedroni (1999), employs a between-dimension approach and comprises three statistics: the group ρ-statistic, the group PP-statistic, and the group ADF-statistic. These statistics are based on estimators that calculate the averages of individually estimated coefficients for each member (Chang & Lee, 2010).

In all seven tests described by Pedroni (1999), the test statistics follow standard normal asymptotic distributions. To achieve this, a standardization process based on the moments of the underlying Brownian motion function is applied (Lee et al., 2008). In particular, the panel v-statistic, a one-sided test, yields significant positive values that result in the rejection of the null hypothesis of no cointegration. On the other hand, the remaining statistics tend to diverge and become infinitely negative, implying that substantial negative values reject the null hypothesis (Lee et al., 2008).

Furthermore, to address endogeneity and serial correlation concerns with the regressors, we will employ the Fully Modified Ordinary Least Squares (FMOLS) estimators as recommended by Pedroni (2000). We will also utilize the Dynamic Ordinary Least Squares (DOLS) approach, which differs in the way observations are combined. Following Pedroni (2000) guidelines, which emphasize the flexibility of Group-means estimators regarding heteroscedasticity in the cointegration vectors and their better size distortion characteristics, we will use this version for both the FMOLS and DOLS estimators (Madeira et al., 2021).

In this study, the estimators used were Pooled Mean Group (PMG), Dynamic Fixed Effect (DFE), Dynamic OLS (DOLS) and Fully Modified OLS (FMOLS). The sample used for this study does not qualify for the Mean Group (MG) estimator, so this estimator was not performed. The OLS estimation of the DOLS aims to solve the problems of other estimators. In addition to being super-consistent, this estimator is also efficient. On the other hand, with the estimation of the FMOLS model), semi-parametric corrections in two steps were made to deal with the problems of auto-correlation of errors.

7.3.4 Pooled Mean Group (PMG), Dynamic Fixed Effect (DFE), Dynamic OLS (DOLS), and Fully Modified OLS (FMOLS)

The PMG approach proposed by Pesaran et al. (1997, 1999) combines pooling and averaging. This combination allows the intercept, the short-term coefficients, and, the term of the error variance to differ between the groups, as occurs in the mean group, however, controls the long-run coefficients to be equal in the entire groups, as occurs in Fixed Effects (Blackburne & Frank, 2007).

The maximum likelihood method was created by Pesaran et al. (1999) so that it is possible to project the parameters. Compelling the likelihood as a multiplication factor for all cross-section's likelihood and applying the log output (Blackburne & Frank, 2007), we have

$$l_T(\theta', \varphi', \sigma') = -\frac{T}{2}\sum_{i=1}^{N} \ln\left(2\pi\sigma_i^2\right) - \frac{1}{2}\sum_{i=1}^{N} \times \frac{1}{\sigma_i^2}\{\Delta y_i - \varnothing_i \xi_i(\theta)\}' H_i \{\Delta y_i - \varnothing_i \xi_i(\theta)\} \tag{7.5}$$

In which, $\xi_i(\theta) = y_{i,\,t\,-\,1} - X_i\theta_i$; $H_i = I_T - W_i\left(W_i' W_i\right) W_i$; I_T is an identity matrix and $W_i = (\Delta y_{i,\,t\,-\,1}, \ldots \Delta y_{i,\,t\,-\,p\,+\,1}, \Delta X_i, \Delta X_{i,\,t\,-\,1}, \ldots \Delta X_{i,\,t\,-\,q\,+\,1})$ (Blackburne & Frank, 2007).

Estimates of parameters of the maximized conditional likelihood iterated are asymptotically identical to those of the maximum likelihood of complete information. According to Blackburne and Frank (2007), the parameters of the Mean Group (MG) approach are the unweighted means of each coefficient.

In cointegration and estimation terms, the DFE is similar to a PMG estimation, in the long-term, dynamic fixed effects need to be the same across all the panels, as well as the pooled mean group. The DFE also limits the speed of the adjustment coefficient and the short-term coefficient to be homogeneous (Altıntaş & Kassouri, 2020), which the PMG does not do. Still, this approach can produce skewed results unless the coefficients are identical.

Once the existence of cointegration is established, the subsequent step involves estimating the corresponding long-run cointegration parameters through two methods: Fully Modified OLS (FMOLS) and Dynamic OLS (DOLS) procedures. Utilization of FMOLS and DOLS is contingent upon the demonstration of cointegration among the series.

DOLS, a parametric approach entails error adjustment through augmentation of the static regression with lagged, leading, and concurrent values of the regressor in the context of first differences (Kurniawati, 2020). In contrast, FMOLS, a nonparametric approach, considers the possible correlation between the error term and the first differences of the regressor (Kurniawati, 2022). It also accommodates the inclusion of a constant term to rectify serial correlation. Furthermore, the FMOLS panel technique tackles the issues related to intercept and endogeneity (Kurniawati, 2022).

If the variables are not cointegrated, the long-term coefficients will be estimated through the FMOLS methodology. The FMOLS edges are correct endogeneity and serial correlation, like the PMG is asymptotically, and eliminates the sample bias. However, it is necessary two conditions, only one integrating vector and the variables cannot be cointegrated (Altıntaş & Kassouri, 2020). According to the model:

$$y_t = \sigma_0 + \sigma_1' X_t + \mu_t, t = 1, 2, \ldots, n \tag{7.6}$$

where y_t is an I(1) variable and X_t is a ($kx1$) vector of I(1) regressors, that aren't cointegrated.

Finally, the DOLS method corrects for potential simultaneity bias among regressors by regressing one of the integrated variables (I(1)) on other integrated variables, alongside the stationary variables (I(0)), and incorporating lags and leads of the first differences of the integrated variables (Chung & Ariff, 2016). Dynamic OLS assumes a particular a priori normalization (Narayan & Narayan, 2005). DOLS can be estimated through the equations below:

$$\Delta X_t^1 = k_t^1 \tag{7.7}$$

$$X_t^2 = \phi_0 + \phi X_t^1 + k_t^1 \tag{7.8}$$

where $X_t' =$, the dimension of X_t^1 and X_t^2 being $(p - r)x1$ and $(rx1)$, in other words, these two matrices have different numbers of rows, which are combine to form the large matrice (Madeira et al., 2021). The error processes are presumed to be stationary, and this is achieved by incorporating both leads and lags of the variables ΔX_t^1 in Eq. (7.8), the normalized cointegrating vectors are estimated ϕ, by OLS, then it is possible to achieve an estimator that is asymptotically similar to Maximum Likelihood Estimation (MLE) (Narayan & Narayan, 2005).

7.3.5 *Data*

To conduct this study on the determinants of economic growth, we utilized a sample comprising 27 European Union countries over the period 1995–2019. We aimed to assess both the long-term effects on economic growth and its determinants, as well as the short-term fluctuations (economic cycles). Economic growth was quantified as the growth rate of natural products. We collected annual observations for several variables: GDP per capita in euros, Carbon Dioxide Emissions (CO_2 per capita) in tonnes, Perception of Corruption Index (where a higher index signifies lower economic growth), Rents of Natural Resources as a percentage of GDP (comprising income from oil, natural gas, coal, minerals, and forests), the Jure Economic Globalization Index, the De Jure Social Globalization Index, and the De Jure Political Globalization Index. All variables underwent natural logarithm transformation, and data were sourced from "Pordata," "Transparency International," Databank, and the "KOF Swiss Economic Institute."

7.4 Results and Discussion

For a comprehensive understanding of the objective outlined in this study, it is essential to delineate the four tangible variables to be incorporated into the analysis concerning economic growth (measured by GDP per capita), Rents of Natural Resources (encompassing rents from natural gas, oil, coal, minerals, and forests), the Transparency of Corruption Index, and the Globalization Index. To grasp certain characteristics of the series within our sample and the cross-sections, it becomes

Table 7.1 Specification tests

	Pooled		Fixed effects	Random effects
Pesaran's test			16.657*** (0.000)	18.139*** (0.000)
Frees' test			4.225	3.944
Friedman's test			126.602*** (0.000)	138.894*** (0.000)
Modified Walf's test			1716.06***	
Wooldridge's test	627.982*** (0.000)		(0.000)	
Pesaran's test	Frees' test	Friedman's test	Modified Walf's test	Wooldridge's test
Reject the null hypothesis	Non-rejection of null hypothesis	Reject the null hypothesis	Reject the null hypothesis	Reject the null hypothesis

Note: The Pesaran, Frees, and Friedman test to test the null cross-section independence hypothesis (cross-section). The Pesaran and Friedman tests follow a standard distribution. The Frees test uses a Q distribution of Frees. The Walf Modified test tests the null hypothesis that the variance of the error term (residues) is constant. The Wooldridge test tests the null hypothesis of a serial correlation, i.e., there is no autocorrelation in the model. The asterisk ***, **, * denote significance of 1%, 5%, and 10%, respectively

imperative to scrutinize the presence of cross-sectional dependence and the order of integration for all variables.

In assessing the existence of cross-sectional dependence among the variables, the Pesaran CD Test was employed. The null hypothesis of this test assumes cross-section independence and was utilized to determine if these variables share common impacts. Furthermore, the Maddala & Wu test, with lag length (1) and variations both without and with trends, was employed.

In the initial phase of the analysis, we employed the Panel-Corrected Standard Errors (PCSE) method for estimation, accompanied by a thorough examination of diagnostic tests (as indicated in Table 7.1). The values of the Pesaran and Friedman tests were found to be lower than the significance levels of 1% and 5%. Consequently, the null hypothesis of cross-sectional independence was rejected, signifying statistical evidence at a 1% significance level, affirming the existence of cross-sectional dependency in the data model.

The results show a contemporary correlation problem, that is, the observations of certain countries are correlated with the observations of others in the same period of time. On the other hand, according to the Frees test, the values indicate that the null hypothesis of cross-sectional independence is not rejected. There is sufficient statistical evidence at the significance level of 1% to conclude that there is no cross-section dependency in the data of this model, and therefore, there are no contemporary correlation problems between the European countries of the sample.

The results of Table 7.2, according to the PCSE's estimators for the European panel, demonstrate that all variables are statistically significant for the three levels of significance (1%, 5%, and 10%), except in the PSAR model (1) and AR (1), where

Table 7.2 PCSE estimator

	CORR (IND)	PSAR (1)	AR (1)	HET—AR (1)
Ln CO_2pc	0.1953355***	0.1594726***	0.2059474***	0.1953355***
Ln perception of corruption	0.3245963***	0.2130058***	0.2129009***	0.3245963***
Ln rents of natural resources (gas, oil, coal, minerals, and forests)	– 0.061585***	– 0.0167878**	– 0.0199169**	– 0.061585***
Ln KOF economic de jure	0.4198882***	0.162488**	0.1645196**	0.4198882***
Ln KOF social de jure	2.482508***	1.778855***	1.767992***	2.482508***
Ln KOF politic de jure	0.6959222***	0.583101***	0.7220449***	0.6959222***
Constant	−7.826488	−2.51601	−3.150621	−7.826488
Obs.	675	675	675	675
R square^2	0.8476	0.9971	0.9899	0.8476
Wald Chi^2	3350.80	288.47	219.78	3132.11

Note: The specifications considered in the empirical analysis are denoted as follows: CORR (IND): Contemporaneous correlation PCSE model. PSAR (1): First-order autocorrelation with a panel-specific AR process coefficient (1). AR (1): PCSE estimator with an option for first-order serial correlation. HET-AR (1): PCSE estimator with options for heteroskedasticity and first-order serial correlation. The significance levels are denoted by asterisks: *** for 1%, ** for 5%, and * for 10% significance

the variables Ln Rents of Natural Resources and Ln KOF Economic de Jure are only significant for 1% and 5%. It can also be concluded that all variables positively influence economic growth, except natural resources that harm growth.

The second generation unit root test was conducted under the null hypothesis (H0) that the variables possess a unit root. The rejection of H0 implies that the variables are stationary, indicating a mix of stationary variables at the level (I(0)) and integrated variables of order one (I(1)). Moreover, the absence of integrated variables of order two (I(2)) was affirmed.

In the second phase of analysis, we studied the relationship between economic growth and the drivers considered, from a dynamic perspective, using the econometric technique of cointegration analysis. We estimated the first generation Pesaran test to conclude whether or not there is cross-sectional dependency. According to the results in Table 7.3, the *P*-values are lower than the significance levels of 1% and 5%, so it is appropriate to reject the null hypothesis of the existence of a unitary root. We also concluded that there is cross-sectional dependency, so it is necessary to perform the second generation of the Pesaran test. The results of this last test, "The Pesaran test (CIPS), assumes that the cross-sectional dependency is in the form of a single, unobserved, common factor," so according to the results and the values of the *P*-value, there is sufficient statistical significance to conclude there is no cross-sectional dependency.

According to Table 7.4, the *P*-value values of the two tests performed are higher than the significance levels of 1% and 5%, respectively, so the null hypothesis of the nonexistence of cointegration is rejected. Therefore, there is enough statistical

Table 7.3 Pesaran's test (CIPS) panel unit-root test results

Variable	Lags	Without trend	With trend
Ln GDP pc	Level (0) first Difference (1)	−1.130 −4.092***	0.272 −3.016***
Ln CO_2 pc	Level (0) first Difference (1)	0.703 −0.419	−0.343 0.955
Ln perception of corruption	Level (0) first Difference (1)	−4.381*** −5.142***	−2.168** −2.861***
Ln rents of natural resources	Level (0) first Difference (1)	−2.150** 0.049	−1.143 0.893
Ln KOF economic de jure	Level (0) first Difference (1)	−2.950*** −1.250	1.897 3.588
Ln KOF social de jure	Level (0) first Difference (1)	−4.973*** −4.985***	−3.561*** −4.180***
Ln KOF politic de jure	Level (0) first Difference (1)	−6.448*** −3.265***	−5.593*** −0.980

Note: The asterisk ***, **, * denote significance of 1%, 5%, and 10%, respectively

Table 7.4 Pedroni's test

	Statistics	*P*-value
Modified Phillips–Perron *t*	5.4972***	0.0000
Augmented dickey-fuller *t*	1.9846**	0.0236

Note: The asterisk ***, **, * denote the significance of 1% and, 5% respectively

evidence with 99% and or 95% confidence to conclude that all panels are cointegrated, so there is a long-term relationship of cointegration.

According to the results presented in Table 7.5, it is evident that the number of variables exhibiting statistical significance in the short term remains nearly consistent in terms of both effect and significance in the long term. Notably, the variables Ln CO_2pc and Ln Perception of corruption consistently maintain their significance both in the short and long term. However, it is worth mentioning that in the DOLS estimator and DFE, the variable Ln Perception of corruption ceases to be significant. Conversely, the variable "Rents of natural resources" does not demonstrate significance in the short term across any estimators but becomes significant in the PMG and DFE estimators in the long term. Furthermore, the variables associated with the three dimensions of globalization (economic, social, and political de jure) display minimal significance in the short term but begin to exhibit greater significance in the long term.

In the long term, the variable "CO_2 emissions" demonstrates a positive and significant impact on GDP per capita in the panel of European countries. The estimation suggests that a 1% increase in CO_2 emissions per capita leads to a 0.2576% increase in economic growth, at a 1% significance level, holding other factors constant. Similarly, a 1% increase in the Perception of Corruption Index is estimated to result in a 0.1340% increase in economic growth (GDPpc), at a 1% significance level, ceteris paribus.

Table 7.5 Estimator results DOLS, FMOLS, PMG e DFE

Variables	FMOLS	DOLS	PMG	DFE
Ln CO_2pc	0.25762***	0.27546***	0.3460***	0.248104***
Ln perception of corruption	0.13400***	0.06403***	0.10431***	0.066364***
Ln rents of natural resources	0.003209	0.005111	0.01066	−0.000315
Ln KOF economic de jure	0.07521***	0.053071	0.107147	0.032951
Ln KOF social de jure	0.153534*	0.206927**	0.223440	0.237295**
Ln KOF political de jure	−0.066914	−0.038568	0.51176***	0.016455
Constant	0.000163	−0.097132	−0.481782**	−0.252944*
ECT			−0.038018**	−0.074037***
D.Ln CO_2pc (−1)	0.01293***	0.01082**	2.07297***	−0.041079
D.Ln perception of corruption (−1)	0.014314**	0.006743	0.57008***	0.095995
D.Ln rents of natural resources (−1)	−0.0013	−0.000773	−0.14654***	−0.189514***
D.Ln KOF economic de jure (−1)	0.04662***	0.052029***	0.499268**	0.57704***
D.Ln KOF social de jure (−1)	0.07003***	0.098854***	6.87834***	2.80615***
D.Ln KOF political de jure (−1)	−0.040945.	−0.0366**	−3.54519***	−0.367303

Note: **, **, * denote significance of 1%, 5%, and 10%, respectively

In terms of short-term effects, if there is an individual acceleration of CO_2 emissions in *t*-1 and an individual acceleration of the corruption perception index, it leads to an acceleration of GDP per capita in year *t*. These effects are statistically significant at the 1% and 5% significance levels respectively. Additionally, upon analyzing the results from the PMG and DFE models, it is observed that the coefficient associated with the error correction term (ECT (−1)) is negative and statistically significant at the 1% significance level. This signifies that any short-term imbalance between the dependent variable (GDP pc) and the explanatory variables converges to the long-term equilibrium relationship. Specifically, the estimate of the error correction term indicates that about 3.80% of the deviation from long-term equilibrium due to a shock in the previous year is corrected in the current year, implying that 7.40% of the imbalances return to the long-term equilibrium.

7.5 Conclusion

The concept of economic growth is restricted and focuses on the quantitative increase of productive capacity, and not on the qualitative transformation of the structure of the economy. The theory of economic growth seeks to find the determining factors of the rate of it and to identify policies that foster its increase. This work estimated the effects of the main determinants of economic growth in the countries of the European Union.

In terms of the empirical methodology employed, the analysis of the panel variables' stationarity involved cross-section dependence tests and first and second generation unit root tests. The presence of cross-section dependence was confirmed, and subsequent robust second generation CIPS tests confirmed that the majority of the variables at the level and in the first differences were integrated of order one, I(1), indicating non-stationarity in the series. With this condition established, Pedroni cointegration tests were conducted to confirm the existence of cointegration. These tests rejected the cointegration hypothesis for most of the European country groups in the sample at both 1% and 5% significance levels. Subsequently, estimations for the two identified long-term relationships were conducted under these cointegration conditions. Different heterogeneous panel estimation methods were employed, including Pooled Mean Group (PMG), Mean Group (MG), Dynamic Fixed Effect (DFE), Dynamic Ordinary Least Square (DOLS), and Fully Modified Ordinary Least Square (FMOLS).

In conclusion, we can say that our results point to the following significant statistical evidence; the variables CO_2 emissions per capita, de jure Corruption and Globalization Index (economic and social de jure) always yield positive results. However, according to theory, a negative influence of CO_2 emissions is expected, as the most industrialized countries, such as European countries, have more CO_2 emissions and economic growth positively influences CO_2 emissions and, in response, CO_2 emissions negatively cause economic growth. The estimated coefficient for the variable, total rents of natural resources (natural gas, oil, coal, minerals, and forests) shows a negative influence on the economic growth of countries. The variable globalization political de jure shows values with significant coefficients in some of the estimators.

In terms of future research, we can propose this line of investigation with the specific aim of analyzing the mediating and interactive effect between environmental policies and renewable energy consumption and consequently, this joint impact on economic growth, to guarantee sustainable economic development. This future research aims to contribute to the current debate on the role of renewable energies by providing evidence of the role of environmental policies in the process of economic growth.

Acknowledgments The authors extend their gratitude to NECE-UBI, Research Unit in Business Science and Economics, sponsored by the Portuguese Foundation for the Development of Science and Technology, project UIDB/04630/2020, for the financial support provided for this study. It is part of the project: PRIMA/0008/2020, which received the DOI 10.54499/PRIMA/0008/2020 (https://doi.org/10.54499/PRIMA/0008/2020).

References

Aidt, T. S. (2009). Institutions and economic development. *Institutions, Transition Economies, and Economic Development, 25*, 113–126. https://doi.org/10.4324/9780429499760-9

Akadiri, S. S., Alkawfi, M. M., Uğural, S., & Akadiri, A. C. (2019). Towards achieving environmental sustainability target in Italy. The role of energy, real income and globalization. *Science of the Total Environment, 671*, 1293–1301. https://doi.org/10.1016/j.scitotenv.2019.03.448

Altıntaş, H., & Kassouri, Y. (2020). Is the environmental Kuznets curve in Europe related to the per-capita ecological footprint or CO2 emissions? *Ecological Indicators, 113*(September), 106187. https://doi.org/10.1016/j.ecolind.2020.106187

Amini, A. (2018). Studying the effect of abundance of natural resources on economic growth. *European Journal of Sustainable Development, 7*(1), 201–208. https://doi.org/10.14207/ejsd.2018.v7n1p201

Amorim Janielly, D. O. (2018). *Corrupção e crescimento econômico.*

Anyanwu, U. M., Anyanwu, A. A., & Cieślik, A. (2021). Does abundant natural resources amplify the negative impact of income inequality on economic growth? *Resources Policy, 74*(March), 1–11. https://doi.org/10.1016/j.resourpol.2021.102229

Ayomitunde, A. T., Akin George, O., Ojo, L. B., & Johnson Ifeanyi, O. (2020). Globalization and economic growth: Evidence from European countries. *European Financial and Accounting Journal, 15*(1), 67–82. https://doi.org/10.18267/j.efaj.234

Baek, J., Cho, Y., & Koo, W. W. (2009). The environmental consequences of globalization: A country-specific time-series analysis. *Ecological Economics, 68*(8–9), 2255–2264. https://doi.org/10.1016/j.ecolecon.2009.02.021

Bataka, H. (2019). De jure, De facto globalization and economic growth in sub-Saharan Africa. *Journal of Economic Integration, 34*(1), 1–9. https://doi.org/10.1037/0033-2909.I26.1.78

Bilgili, F., & Ulucak, R. (2020). Does globalization matter for environmental sustainability? Empirical investigation for Turkey by Markov regime switching models. *Environmental Science and Pollution Research, 27*, 1087–1100.

Blackburne, E. F., & Frank, M. W. (2007). Estimation of nonstationary heterogeneous panels. *The Stata Journal, 1*(3), 1–22.

Borensztein, E., De Gregorio, J., & Lee, J.-W. (2005). How does foreign direct investment affect economic growth? *Motricite Cerebrale, 26*(2), 59–63. https://doi.org/10.1016/S0245-5919(05)81667-X

Chang, C. P., & Lee, C. C. (2010). Globalization and economic growth: A political economy analysis for OECD countries. *Global Economic Review, 39*(2), 151–173. https://doi.org/10.1080/1226508X.2010.483835

Chung, T. F., & Ariff, M. (2016). A test of the linkage among money supply, liquidity and share prices in Asia. *Japan and the World Economy, 39*, 48–61. https://doi.org/10.1016/j.japwor.2016.07.001

d'Agostino, G., Dunne, J. P., & Pieroni, L. (2016). Corruption and growth in Africa. *European Journal of Political Economy, 43*, 71–88. https://doi.org/10.1016/j.ejpoleco.2016.03.002

Danish, Ulucak, R., & Khan, S. U. D. (2020). Relationship between energy intensity and CO2 emissions: Does economic policy matter? *Sustainable Development, 28*(5), 1457–1464. https://doi.org/10.1002/sd.2098

Djankov, S., La Porta, R., Lopez-de-Silanes, F., & Shleifer, A. (2002). The regulation of entry. *Quarterly Journal of Economics, 117*(1), 1–37. https://doi.org/10.1162/003355302753399436

Dreher, A. (2006). Does globalization affect growth? Evidence from a new index of globalization. *Applied Economics, 38*(10), 1091–1110. https://doi.org/10.1080/00036840500392078

Dreher, A., & Gassebner, M. (2013). Greasing the wheels? The impact of regulations and corruption on firm entry. *Public Choice, 155*(3–4), 413–432. https://doi.org/10.1007/s11127-011-9871-2

Egbetunde, T., & Akinlo, A. E. (2015). Financial globalization and economic growth in sub-Saharan Africa: Evidence from panel cointegration tests. *African Development Review, 27*(3), 187–198. https://doi.org/10.1111/1467-8268.12140

Frankel, J. (2010). *The natural resource curse: A survey*. National Bureau of Economic Research.

European Commission. (n.d.). The European Green Deal. *European Commission*. https://commission.europa.eu/strategy-and-policy/priorities-2019-2024/story-von-der-leyen-commission/europeangreen-deal_en

Gasimli, O., Haq, I. u., Munir, S., Khalid, M. H., Gamage, S. K. N., Khan, A., & Ishtiaq, M. (2022). Globalization and sustainable development: Empirical evidence from CIS countries. *Sustainability (Switzerland), 14*(22), 1–15. https://doi.org/10.3390/su142214684

González-Álvarez, M. A., & Montañés, A. (2023). CO2 emissions, energy consumption, and economic growth: Determining the stability of the 3E relationship. *Economic Modelling, 121*(February), 106195. https://doi.org/10.1016/j.econmod.2023.106195

González-Val, R., & Pueyo, F. (2019). Natural resources, economic growth and geography. *Economic Modelling, 83*(July 2018), 150–159. https://doi.org/10.1016/j.econmod.2019.02.007

Gozgor, G., Kumar, M., Demir, E., & Padhan, H. (2020). The impact of economic globalization on renewable energy in the OECD countries. *Energy Policy, 139*(June), 111365. https://doi.org/10.1016/j.enpol.2020.111365

Guan, J., Kirikkaleli, D., Bibi, A., & Zhang, W. (2020). Natural resources rents nexus with financial development in the presence of globalization: Is the "resource curse" exist or myth? *Resources Policy, 66*(February), 101641. https://doi.org/10.1016/j.resourpol.2020.101641

Gyamfi, B. A., Bein, M. A., & Bekun, F. V. (2020). Investigating the nexus between hydroelectricity energy, renewable energy, nonrenewable energy consumption on output: Evidence from E7 countries. *Environmental Science and Pollution Research, 27*(20), 25327–25339. https://doi.org/10.1007/s11356-020-08909-8

Gygli, S., Haelg, F., Potrafke, N., & Sturm, J. E. (2019). The KOF globalisation index—Revisited. *Review of International Organizations, 14*(3), 543–574. https://doi.org/10.1007/s11558-019-09344-2

Haini, H., Wei Loon, P., Yong, S. K., & Husseini, S. (2023). Does social globalization affect the relationship between international tourism and economic growth? *Journal of Travel Research, 63*(1), 252–269. https://doi.org/10.1177/00472875221146779

Haseeb, A., Xia, E., Danish, Baloch, M. A., & Abbas, K. (2018). Financial development, globalization, and CO2 emission in the presence of EKC: Evidence from BRICS countries. *Environmental Science and Pollution Research, 25*(31), 31283–31296. https://doi.org/10.1007/s11356-018-3034-7

Huisman, R., & Kiliç, M. (2015). Time variation in European carbon pass-through rates in electricity futures prices. *Energy Policy, 86*, 239–249. https://doi.org/10.1016/j.enpol.2015.07.005

Huntington, S. P. (1968). Political order in changing societies. In *Political order in changing societies*. https://doi.org/10.5771/0506-7286-1970-2-257

Kaufmann, D. (1997). *Corruption: The facts*. pp. 1–17.

Khaliq, A., & Noy, I. (2007). Foreign direct investment and economic growth: Empirical evidence from sectoral data in Indonesia. *Journal of Economic Literature, 45*(1), 313–325.

Khumawala, S., & Ramchand, L. (2005). Country level corruption and frequency of issue in the U.S. market. *Journal of Public Budgeting, Accounting & Financial Management, 17*, 341–364.

Kurniawati, M. A. (2020). The role of ICT infrastructure, innovation and globalization on economic growth in OECD countries, 1996-2017. *Journal of Science and Technology Policy Management, 11*(2), 203–225. https://doi.org/10.1108/JSTPM-06-2019-0065

Kurniawati, M. A. (2022). Analysis of the impact of information communication technology on economic growth: Empirical evidence from Asian countries. *Journal of Asian Business and Economic Studies, 29*(1), 2–18. https://doi.org/10.1108/jabes-07-2020-0082

Leal, P. H., Marques, A. C., & Shahbaz, M. (2020). The role of globalisation, de jure and de facto, on environmental performance: Evidence from developing and developed countries. *Environment, Development and Sustainability, 23*, 7412. https://doi.org/10.1007/s10668-020-00923-7

Lee, C., Chang, C., & Chen, P. (2008). Energy-income causality in OECD countries revisited: The key role of capital stock. *Energy Economics, 30*, 2359–2373. https://doi.org/10.1016/j.eneco.2008.01.005

Leff, N. H. (1964). Economic development through bureaucratic corruption. *American Behavioral Scientist, 8*(3), 8–14. https://doi.org/10.1177/000276426400800303

Li, S., & Wu, J. (2010). Why some countries thrive despite corruption: The role of trust in the corruption-efficiency relationship. *Review of International Political Economy, 17*(1), 129–154. https://doi.org/10.1080/09692290802577446

Liu, M., Ren, X., Cheng, C., & Wang, Z. (2020). The role of globalization in CO2 emissions: A semi-parametric panel data analysis for G7. *Science of the Total Environment, 718*, 137379. https://doi.org/10.1016/j.scitotenv.2020.137379

Madeira, A., Moutinho, V., & Alberto, J. (2021). Does waiting times decrease or increase operational costs in short and long-term? Evidence from Portuguese public hospitals. *The European Journal of Health Economics, 22*(8), 1195–1216. https://doi.org/10.1007/s10198-021-01331-y

Majidi, A. F. (2017). Foreign trade, human capital and economic growth: Evidence from Asian countries. *International Journal of Asian Social Science, 7*(12), 942–948. https://doi.org/10.18488/journal.1.2017.712.942.948

Martens, P., Caselli, M., De Lombaerde, P., Figge, L., & Scholte, J. A. (2015). New directions in globalization indices. *Globalizations, 12*(2), 217–228. https://doi.org/10.1080/14747731.2014.944336

Mauro, P. (1995). Corruption and growth. *The Quarterly Journal of Economics, 110*(3), 681–712. https://doi.org/10.2298/fid1301021m

Mauro, P. (2004). The persistence of corruption and slow economic growth. *IMF Staff Papers, 51*(1), 1–18.

Mert, M., & Bölük, G. (2016). Do foreign direct investment and renewable energy consumption affect the CO2 emissions? New evidence from a panel ARDL approach to Kyoto annex countries. *Environmental Science and Pollution Research, 23*(21), 21669–21681. https://doi.org/10.1007/s11356-016-7413-7

Moghaddam, A. A., & Redzuan, M. (2012). Globalization and economic growth: A case study in a few developing countries (1980-2010). *Research in World Economy, 3*(1), 54–62. https://doi.org/10.5430/rwe.v3n1p54

Moiseev, N., Mikhaylov, A., Varyash, I., & Saqib, A. (2020). Investigating the relation of gdp per capita and corruption index. *Entrepreneurship and Sustainability Issues, 8*(1), 780–794. https://doi.org/10.9770/jesi.2020.8.1(52)

Murphy, K. M., Shleifer, A., & Vishny, R. W. (1991). The allocation of talent: Implications for growth. *The Quarterly Journal of Economics, 106*(2), 503. https://doi.org/10.2307/2937945

Musila, J. W., & Yiheyis, Z. (2015). The impact of trade openness on growth: The case of Kenya. *Journal of Policy Modeling, 37*(2), 342–354. https://doi.org/10.1016/j.jpolmod.2014.12.001

Narayan, P. K., & Narayan, S. (2005). Estimating income and price elasticities of imports for Fiji in a cointegration framework. *Economic Modelling, 22*(3), 423–438. https://doi.org/10.1016/j.econmod.2004.06.004

Olimpia, N., & Stela, D. (2017). Impact of globalisation on economic growth in Romania: An empirical analysis of its economic, social and political dimensions. *Studia Universitatis "Vasile Goldis" Arad—Economics Series, 27*(1), 29–40. https://doi.org/10.1515/sues-2017-0003

Pata, U. K., & Yilanci, V. (2020). Financial development, globalization and ecological footprint in G7: Further evidence from threshold cointegration and fractional frequency causality tests. *Environmental and Ecological Statistics, 27*(4), 803–825. https://doi.org/10.1007/s10651-020-00467-z

Pedroni, P. (1999). Critical values for cointegration tests in heterogeneous panels with multiple of economics and statistics. *Oxford Bulletin of Economics and Statistics, 61*, 653–670.

Pedroni, P. (2000). *Fully modified OLS for heterogeneous cointegrated.* Emerald Group Publishing Limited.

Pejović, B., Karadžić, V., Dragašević, Z., & Backović, T. (2021). Economic growth, energy consumption and CO2 emissions in the countries of the European Union and the Western Balkans. *Energy Reports, 7*, 2775–2783. https://doi.org/10.1016/j.egyr.2021.05.011

Peng, H., Tan, X., Li, Y., & Hu, L. (2016). Economic growth, foreign direct investment and CO2 emissions in China: A panel granger causality analysis. *Sustainability (Switzerland), 8*(3), 233. https://doi.org/10.3390/su8030233

Pesaran, M. H. (2007). A simple panel unit root test in the presence of cross-section dependence. *Journal of Applied Econometrics, 22*(2), 265–312. https://doi.org/10.1002/jae.951

Pesaran, M. H., Pesaran, M. H., Shin, Y., & Smith, R. P. (1999). Pooled mean group estimation of dynamic heterogeneous panels. *Journal of the American Statistical Association, 94*(446), 621–634. https://doi.org/10.1080/01621459.1999.10474156

Pesaran, M. H., Shin, Y., & Smith, R. P. (1997). Pooled estimation of long-run relationship in dynamic heterogeneous panels.

Radulović, M., & Kostić, M. (2020). Globalization and economic growth of eurozone economies. *Zbornik Radova Ekonomskog Fakultet Au Rijeci, 38*(1), 161–192. https://doi.org/10.18045/zbefri.2020.183

Rahim, S., Murshed, M., Umarbeyli, S., Kirikkaleli, D., Ahmad, M., Tufail, M., & Wahab, S. (2021). Do natural resources abundance and human capital development promote economic growth? A study on the resource curse hypothesis in next eleven countries. *Resources, Environment and Sustainability, 4*(December), 100018. https://doi.org/10.1016/j.resenv.2021.100018

Rahman, M. M. (2020). Environmental degradation: The role of electricity consumption, economic growth and globalisation. *Journal of Environmental Management, 253*(July), 109742. https://doi.org/10.1016/j.jenvman.2019.109742

Samimi, P., & Jenatabadi, H. S. (2014). Globalization and economic growth: Empirical evidence on the role of complementarities. *PLoS One, 9*(4), 1–7. https://doi.org/10.1371/journal.pone.0087824

Santiago, R., Fuinhas, J. A., & Marques, A. C. (2020). The impact of globalization and economic freedom on economic growth: The case of the Latin America and Caribbean countries. *Economic Change and Restructuring, 53*(1), 61–85. https://doi.org/10.1007/s10644-018-9239-4

Sarkodie, S. A., & Strezov, V. (2019). Effect of foreign direct investments, economic development and energy consumption on greenhouse gas emissions in developing countries. *Science of the Total Environment, 646*, 862–871. https://doi.org/10.1016/j.scitotenv.2018.07.365

Sethi, P., Chakrabarti, D., & Bhattacharjee, S. (2020). Globalization, financial development and economic growth: Perils on the environmental sustainability of an emerging economy. *Journal of Policy Modeling, 42*(3), 520–535. https://doi.org/10.1016/j.jpolmod.2020.01.007

Shahbaz, M., Khan, S., Ali, A., & Bhattacharya, M. (2017). The impact of globalization on Co2 emissions in China. *Singapore Economic Review, 62*(4), 929–957. https://doi.org/10.1142/S0217590817400331

Shahbaz, M., Mallick, H., Mahalik, M. K., & Loganathan, N. (2015). Does globalization impede environmental quality in India? *Ecological Indicators, 52*, 379–393. https://doi.org/10.1016/j.ecolind.2014.12.025

Shahbaz, M., Mallick, H., Mahalik, M. K., & Sadorsky, P. (2016). The role of globalization on the recent evolution of energy demand in India: Implications for sustainable development. *Energy Economics, 55*, 52–68. https://doi.org/10.1016/j.eneco.2016.01.013

Shahbaz, M., Shahzad, S. J. H., Mahalik, M. K., & Hammoudeh, S. (2018). Does globalisation worsen environmental quality in developed economies? *Environmental Modeling and Assessment, 23*(2), 141–156. https://doi.org/10.1007/s10666-017-9574-2

Swadźba, S. (2020). Globalization and economic growth in highly developed countries. *SHS Web of Conferences, 74*, 05025. https://doi.org/10.1051/shsconf/20207405025

Uberti, L. J. (2021). Corruption and growth: Historical evidence, 1790–2010. *Journal of Comparative Economics, 50*(2), 321–349. https://doi.org/10.1016/j.jce.2021.10.002

Udemba, E. N., Magazzino, C., & Bekun, F. V. (2020). Modeling the nexus between pollutant emission, energy consumption, foreign direct investment, and economic growth: New insights from China. *Environmental Science and Pollution Research, 27*(15), 17831–17842. https://doi.org/10.1007/s11356-020-08180-x

Usman, O., Akadiri, S. S., & Adeshola, I. (2020). Role of renewable energy and globalization on ecological footprint in the USA: Implications for environmental sustainability. *Environmental Science and Pollution Research, 27*(24), 30681–30693. https://doi.org/10.1007/s11356-020-09170-9

Van Der Ploeg, F. (2011). Natural resources: Curse or blessing? *Journal of Economic Literature, 49*(2), 366–420. https://doi.org/10.1257/jel.49.2.366

Wang, Z., Bui, Q., Zhang, B., & Pham, T. L. H. (2020). Biomass energy production and its impacts on the ecological footprint: An investigation of the G7 countries. *Science of the Total Environment, 743*, 140741. https://doi.org/10.1016/j.scitotenv.2020.140741

Woo, B., & Jun, H. (2020). Globalization and slums: How do economic, political, and social globalization affect slum prevalence? *Habitat International, 98*(August), 102152. https://doi.org/10.1016/j.habitatint.2020.102152

Wooldridge, J. M. (2002). Econometric analysis of cross section and panel data. In *Advanced studies in theoretical and applied econometrics*. https://doi.org/10.1007/978-3-319-60783-2_6

Zahonogo, P. (2018). Globalization and economic growth in developing countries: Evidence from sub-Saharan Africa. *International Trade Journal, 32*(2), 189–208. https://doi.org/10.1080/08853908.2017.1333933

Chapter 8
The Problem of Regional Energy Scarcity in Europe

Yuri Yegorov

Abstract Global energy resources are characterized by substantial spatial heterogeneity across continents, countries and regions. Even in globalized world geographic approach to modelling was important, for example, for natural gas. The failure of globalization and coming regionalization put a pressure for some regions, Europe, in particular. It needs ambitious fast transition to green energy. But even here we observe internal regional heterogeneity related to possibilities in regional production of water, solar and wind energy. Stochastic supply calls for building powerful HV network to balance regional demand with supply, and this requires higher integration inside EU. This chapter discusses this issue from theoretical perspective of energy security after green transition, taking into account both physical and economic issues. The influence of geopolitics on natural gas scarcity is also discussed.

8.1 Introduction

The goal of this chapter is to reveal old problems with energy infrastructure in Europe and the increased importance of it with the recent geopolitical changes and green transition. In a stationary economy (fixed suppliers of energy for decades and no technological changes) existing infrastructure (here high voltage lines for electricity, internal gas pipelines and LNG regasification ports) should solve the problem of flows across regions and have some minimal excess capacity. This extra capacity is costly, but it allows us to solve the security problems in critical moments. The chapter is interdisciplinary; it uses ideas from both physics and economics, but intends to present them in rather simple manner.

When structural changes (like green transition) take place, we have new flows of energy across regions, and new infrastructure capacity should be added. In

Y. Yegorov (✉)
Economica, Vienna, Austria
e-mail: yuri.yegorov@economica.at

© The Author(s), under exclusive license to Springer Nature Switzerland AG 2024
T. Devezas et al. (eds.), *Global Energy Transition and Sustainable Development Challenges, Vol. 1*, World-Systems Evolution and Global Futures,
https://doi.org/10.1007/978-3-031-67583-6_8

particular, this is important for balancing wind energy with itself across regions (see Subsect. 3.1). Some changes emerge randomly as geopolitical consequences. Gas infrastructure inside EU (storage capacities and gas pipelines) depends both on the composition of inflows and changes in regional demand. In general, EU has replaced some of pipeline gas imports by LNG imports, and this requires adapting the capacities of its internal gas pipeline network. This issue will be considered in Sect. 5.

While the transition to green energy in Europe was discussed for almost a decade (after Paris agreement on activities to mitigate climate change), the political situation since 2022 (war in Ukraine and interruption of supply of hydrocarbon from Russia to EU) has created an incentive for EU to do this process fast.

There already exist plans for huge investment in solar and wind energy in Germany and other countries. But building production capacity is not sufficient. Non-storability of electricity in large volumes requires balance of random supply with less random (but heterogeneous) demand both in time and space. And this requires huge capacity for electricity transmission over large distances.

The development of high voltage grid is an important component for energy security. Its insufficient capacity can result in a bottleneck congestion and inability to react on extreme situations like natural disasters. Moving to sustainable energy supply in future requires higher penetration of wind energy with very uncertain and spatially heterogeneous supply. This requires balancing of a wind with the other energy sources. High wind penetration will make this problematic unless wind is partly balanced with itself. This will require regional integration by HV lines at substantial distances (Yegorov & Wirl, 2018).

Section 2 gives basic facts about European energy composition from primary energies. Section 3 presents physical foundations for scarcity of electric energy. Section 4 briefly discusses the role of oil and its supply in EU. Section 5 studies the problem of scarcity of natural gas. Section 6 concludes.

8.2 European Energy Composition

In 2019, primary energy consumption in Europe was at 83.46 exajoules (EJ), while in 2020 it went down by 7.8% to 77.15 EJ, due to covid pandemic. At this time, there was no shortage of supply of all types of energy, but the demand went down (BP, 2021, p. 10). However, energy consumption in EU did not recover after covid19 pandemic, but now the reason was mostly in supply and high prices for electricity and gas since 2021, which were also causing high inflation in EU.

In EU, consumption share of fossil fuels in 2020 was at 71%, below the global share. The share of renewable energy was 12%, hydro 7% and nuclear 9% (BP, 2021).

The consumption of natural gas in Europe in 2020 has dropped only by 2.5% (from 553.5 bcm to 541.1 bcm (billion cubic metres)) (BP, 2021, p. 38). A large share of it (326 bcm) has been imported, mostly as pipeline gas (211.3 bcm), but also in the form of LNG (114.8 bcm) (BP, 2021, p. 42). While the share of imported LNG grew by 3.8% in 2020, it was still far from dominant (BP, 2021, p. 42). In 2020 Russia has supplied 152.1 bcm of pipeline gas and 17.2 bcm of LNG to Europe. The second largest supplier was Norway, with 106.9 bcm of pipeline gas and 4.1 bcm of LNG. 30.2 bcm of LNG came from the Middle East (mostly Qatar), while other came from different countries in Africa and America (BP, 2021, p. 45).

That is why disruption of supply of pipeline gas in 2022 (due to sanctions over Russia and explosion of Nord Stream) had a very important impact on EU natural gas scarcity, at least in the short-term horizon. Most of electricity in EU is generated using renewable energies, while the share of fossil fuels in its generation in 2020 was about 35% (BP, 2021, p. 64). Europe has a fast growth in production and consumption of renewable energy (solar and wind, nuclear and biofuels are counted separately), 8.2% in 2020, but the overall volume is still rather small (8.94 EJ, or 8%) (BP, 2021, p. 55).

The capacity of solar and wind energy in Europe is rather high (167.8 GW for solar and 216.8 GW for wind) and is growing fast (14.4% for solar and 5.0% for wind in 2020). In 2022, European solar capacity grew by 14.7% and wind capacity by 7.7% (BP, 2022). However, the major problem with those energies is in temporal heterogeneity of supply of solar energy and randomness of supply of wind energy. This brings the necessity to balance those energies over space and time. As for temporal balance, high capacity of batteries and accumulators is required, and their cost is still high. As for regional balance, there is a necessity to have high capacity of infrastructure (HV lines for electricity and pipelines for natural gas and future hydrogen). This question will be addressed in Sect. 3 in more detail.

8.3 Physical Foundations of Energy Scarcity

The concept of energy security has several components, including not only economical, but also physical. Contrary to natural gas and oil, electricity cannot be stored in large volumes (only in accumulators). Electricity and gas are important components of primary energies and will be considered separately.

8.3.1 Electricity: Non-Storability and Necessity of HV Lines

At present, low transmission capacity due to lack of HV connections across the EU countries creates bottlenecks for long distance electricity transmission, resulting in a congestion, quotas, high transmission price and finally market regionalization. The EU can invest in a new capacity of HV lines to achieve 2 goals: a) higher market integration, with price convergence, b) more possibilities to balance an increasing share of renewable energies, especially wind electricity that may have low correlation at large distances.

Yegorov and Wirl (2018) have studied this question from a theoretical perspective. In order to balance random wind supply with itself, but in a different location, it is necessary to analyse spatial correlation of wind fields. The wind correlation between two particular points in space can be calculated after aggregation over large time intervals. Correlation function can be constructed when we have a sequence of points on a line at different distances. Typically, it starts from 1 at zero distance and decays with distance. The idea about benefits from interconnection of wind farms is well known, at least among meteorologists. The wind speed correlation among sites decreases when more farms and at larger distance are interconnected.

Toshiaki et al. (2009) studied correlation of wind between closely located meteorological stations in a particular coastal region in Japan. Correlation falls to about 0.6 at distances 20–25 km. However, Japan has very mountainous landscape, which is an obstacle for the wind propagation. Haslett and Raftery (1989) studied correlation as the function of distance for 12 points spread over all Ireland (area about 500x300 km). We can observe almost a linear decay of correlation with the distance. It is about 0.85 for the distance of 100 km and vanishes to 0.6 only at 400 km. These results predict much higher radius of correlation than in Japan.

Archer and Jacobson (2007) did the simulation of wind aggregation for a large area (square of size 850 km) in the USA. It was found that an average of 33% and a maximum of 47% of yearly averaged wind power from interconnected farms can be used as reliable, baseload electric power. This suggests that EU should also balance the wind on the territories of similar size. Since the size of EU country is typically 300–600 km, this will imply aggregation across several countries and the necessity to build substantial HV capacity across countries.

Those studies suggest the necessity to integrate wind for proper balance over large areas, with 500–1000 km in diameter, which exceeds the size of an average EU country. This requires extra capacity for HV lines. Jamasb Tooraj (2013) presents the map of HV lines in EU. It is rather dense and contains mostly AC lines with capacities between 220 and 500 kV as well as DC lines in Baltic Sea, between Denmark, Norway, Sweden, Finland, Germany and Poland.

8.3.2 Justification of Possibility to Use HV Lines for Energy Balance in Europe

Electricity transmission over high distances leads to losses. However, those losses depend on wires and voltage and for HV lines can be acceptable. Such analysis was also provided in Yegorov and Wirl (2018). Typical losses for AC lines are as follows. A 100 miles (160 km) span at 765 kV carrying 1000 MW of power can have losses between 0.5% and 1.1%. A 345 kV line carrying the same load across the same distance has losses of 4.2%, or about 4–8 times higher. If we need to send

Table 8.1 Technical characteristics of HV lines. Source: treugoma.ru

Voltage, kV	Transmitted power, MW	Maximal distance, km
220	135	400
330	360	700
500	900	1200
750	2100	2200
1150	5200	3000

electricity at distances about 1500 km, it is better to use 750 kV lines, because the losses are still below 10%.

Table 8.1 gives the transmitted power (in MW) and the limit length in km (where 10% of power is lost) as the function of voltage (kV), based on information from Treugoma.ru (n.d.).

8.3.3 EU Plans for Decarbonization and the Necessity to Close Bottlenecks

The EU has very ambitious plans for future penetration of wind energy. New installed wind capacity in 2016 exceeded installed capacities of all other types of generation. Between 2000 and 2016, EU has installed 142.6 GW of wind power, followed by 101.2 GW of solar power and only 98.5 GW of natural gas (Wind Europe, 2016). In 2016 overall wind capacity took over coal and is approaching to natural gas. Those installations are quite heterogeneous across EU countries. Germany is the EU leader in both capacity (62.2 GW in 2020) and the speed of new installations (5.44 GW in 2016). Next goes Spain with 27.1 GW of wind capacity in 2020, but its new installations have slowed down. UK occupies the third position in EU with 24.7 GW of wind capacity in 2020.

To ensure that offshore renewable energy can help to reach the EU's ambitious energy and climate targets for 2030 and 2050, the Commission published a dedicated EU strategy on offshore renewable energy (COM (2020) 741) on 19 November 2020. The strategy sets targets for an installed capacity of at least 60 GW of offshore wind and 1 GW of ocean energy by 2030, and 300 GW and 40 GW, respectively, by 2050.

The future EU plans for the decarbonization scenario are even more ambitious. In the Energy Roadmap 2050 (EC) there are targets to increase the share of renewable energy among all (not only power generation) consumed primary energies to 25% in 2030 and to 50% in 2050. Transition to renewable energies will tackle both the problems of import dependency in energy (that is also a part of energy security problem) and climate change (Wirl & Yegorov, 2014).

But is the EU ready for this ambitious task with its HV network? Currently there are many bottlenecks in the EU energy infrastructure (Opitz & Tellman, 2015). Most of these bottlenecks are between price zones, which partly coincides with some

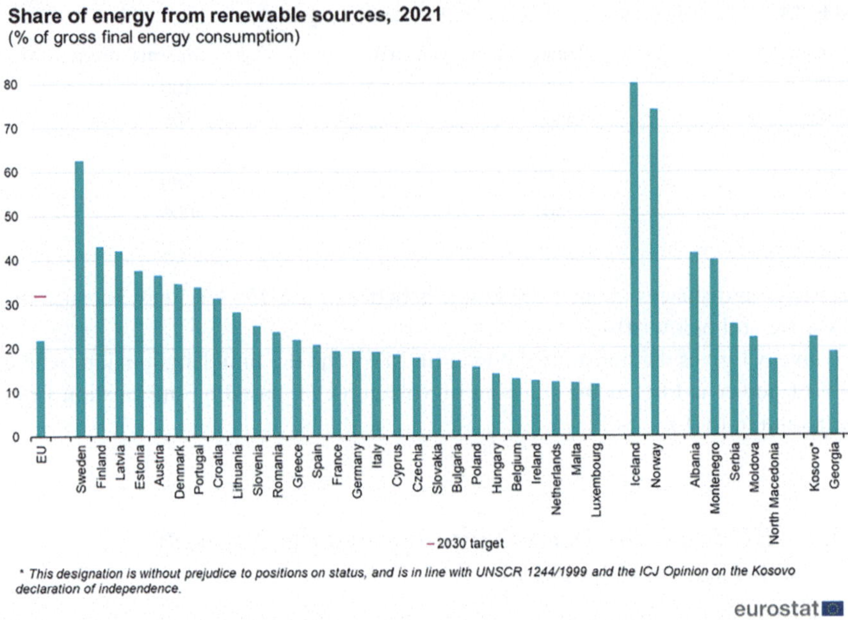

Fig. 8.1 Share of energy from renewable sources, 2021, for EU countries. Source: Eurostat

country borders (like France–Spain, Spain–Portugal). Long countries (Italy, Sweden, Norway) have also bottlenecks inside their territories (between North and South). Germany also has several bottlenecks inside. Those bottlenecks existed under moderate penetration of renewable energies, and demand for new HV lines will increase substantially after transition to green energies. The reason is the necessity to balance random supply of wind energy across regions. According to grid plans from 35 countries, over 25,000 km of new lines are planned between now and 2026, increasing the total length of national transmission networks by 5.3%. This would bring the total length to approximately 523,000 km by the end of 2026. Most of these lines will be constructed in Spain, followed by Germany and Denmark.[1]

The EU reached a 21.8% share of its gross final energy consumption from renewable sources in 2021 (Eurostat. Statistics explained 2021). Share of renewable energy more than doubled between 2004 and 2021. Figure 8.1 shows those shares for different European countries, showing huge heterogeneity. Iceland, Norway and Sweden are the leaders, but many EU countries have the share of renewable energy below 15%. This suggests heterogeneous transition to replacement of fossil fuels by renewables across EU countries, and thus changing flows of energies across them along the transition path.

[1] Source: https://ember-climate.org/insights/research/putting-the-mission-in-transmission-grids-for-europes-energy-transition/

While the EU was originally designed as infrastructure union (for coal and steel industries), its current policy towards full integration of infrastructures is far from that. There is relatively little trade in electricity across countries, and there are many bottlenecks of capacity, which does not allow for this trade. In a stationary environment (before 2020s) the low demand for such trade was regulated by high costs, via auctions for electricity transmission in bottlenecks. On the one hand, this allowed national electricity companies to keep local monopoly (and sometimes charge higher prices from local consumers, comparing to neighbouring countries). But on the other hand, this became a weak point in a new world of green energy, which will require huge transmission of electricity across countries to balance random supply of wind energy.

8.4 Oil

Oil and its products still play the major role in global portfolio of primary energies. According to BP (2020), the share of oil in globally consumed primary energies has dropped from 39% in 1999 to 33% in 2019 and is still above the share of coal (27%) and natural gas (24%). In Europe oil share was 38% in 2019. EU has higher share of renewable energies (11% in 2019 versus 5% globally) and this share is growing, but they (together with hydro and nuclear energy, that are almost stable) cannot replace oil and other fossil fuels substantially in the next 10 years, despite many efforts. Those policies will lead to some decline in oil consumption, but not too fundamental.

According to WEO (2014), in the New Policies Scenario, before 2040 global demand for primary energy will grow on average annual rate of 1.1%, while in EU it will decline by 0.3% per year on average. However, EU depends heavily on imports of oil and gas, having low level of deposits, especially for oil. In EU, import share of oil is expected to rise from 85% to 89%, while for gas—from 66% to 82%. At the same time, the USA will become less dependent on imports. Shale gas revolution even made them net gas exporters at present.

8.5 Natural Gas

Natural gas is still an important fuel not only today (with about 25% of world primary energies) but also in close future. The reason is that it produces less carbon emissions and less pollution comparing to coal. At the same time, very fast transition to renewable energy is impossible, because of high investment, necessity to upgrade technologies and some other problems (of physical, economic and socio-political origin), discussed also in Wirl and Yegorov (2014).

Despite transition to green energy in EU and high prices of natural gas in 2022 (caused by geopolitical reasons), natural gas consumption in Europe did not decline in 2022, but grew at 5.7% (BP, 2022). The consumption of coal has also increased by 5.9% (BP, 2022), although this contradicts the principles of green energy transition.

Supply security depends also on the development of internal infrastructure in a country or region. That includes pipelines to a country's border and regasification plants in its ports. There is a trade-off: lack of capacity to accept the supplied gas will lead to its shortage, while spare capacity involves an additional cost. The absence of a large pipeline between Turkey and the EU prevents the development of a secondary gas infrastructure in Balkan countries (Yegorov, 2018) and becomes risky for land-locked EU countries, without LNG ports.

While the lion share of electricity is produced domestically (cross-country trade was small, but can increase in future due to the necessity to balance demand and supply across regions), most of natural gas in EU was imported. Gas can arrive in two forms, via pipeline and in the form of LNG. Before 2022, about 1/3 of natural gas to EU arrived from Russia via pipelines.

The capacity is an important constraint for moving the natural gas across regions. The real flow is the minimum between mutually agreed trade flow and capacity. Access of a consumer to different producers is also limited, because they are linked via a pipeline grid with certain topology and a capacity limit between each pair of nodes. This capacity limit is determined by technical parameters. In general, higher capacity means higher investment cost. Contrary to electricity, it is possible to store gas. This allows moving misbalance between demand and supply over time.

Any analysis of the gas market must include economic theory, technology, geography and geopolitics. The phenomenon of gas transit has a geopolitical origin and may lead to games (Yegorov & Wirl, 2010).

Liquified natural gas allows to avoid transit games, but it has additional costs: the necessity to build liquefaction and regasification plants. In 2008, Europe was consuming mostly pipeline gas, and it had only 16 regasification plants, mostly located in Spain. In 2020, Europe had more regasification plants, located in Spain (7), France (3), Italy (3), UK (3), Belgium, Netherlands, Greece and Lithuania.[2]

Figure from Marin (2021) shows the map of European infrastructure (pipelines and LNG ports), related to natural gas, in 2020. We observe an increase of LNG ports (10 years ago only Spain was relying mostly on LNGs) and huge dependence on natural gas imports from Russia before 2022 (about 170 bcm as pipeline gas, moving by several pipelines, and 20 bcm of LNG). Marin (2021) presents a map of EU with pipelines and LNG terminals in 2019, along with annual gas consumption by different EU countries (this information is also available in BP Statistical Yearbook). The second largest supplier of gas was Norway, with 109 bcm. The right-bottom picture there shows that before 2020 the pipelines via Ukraine have been the main route, but it was replaced by Nord Stream from January 2020. Since fall 2022 Nord Stream is no longer in operation.

[2]Source: Cécile Marin, May 2021, Le Monde Diplomatique.

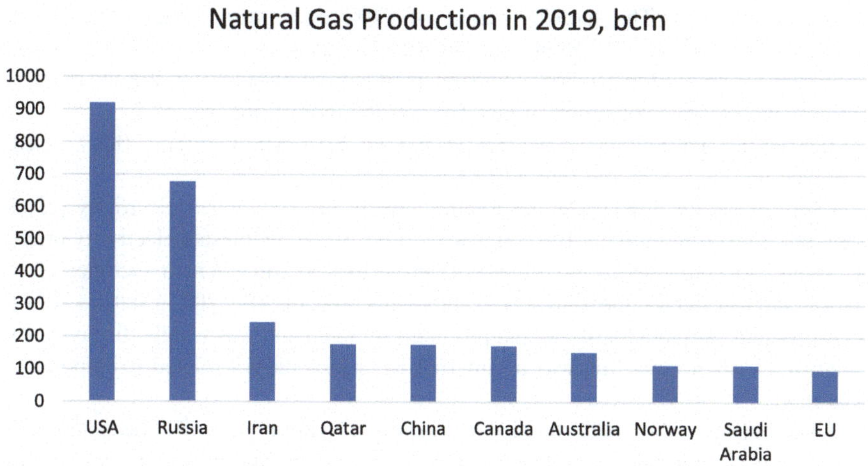

Fig. 8.2 Top 10 producers of natural gas in 2019, in bcm. Source: BP (2020), own elaboration

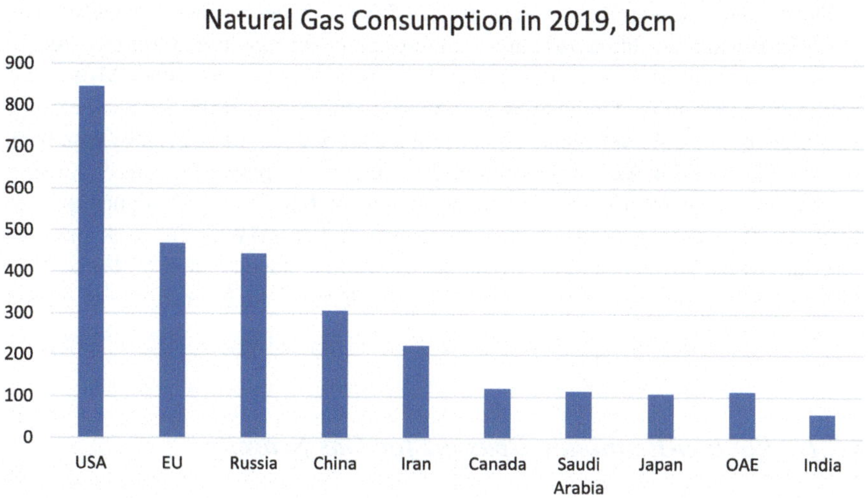

Fig. 8.3 Top 10 consumers of natural gas in 2019, in bcm. Source: BP (2020), own elaboration

Figure 8.2 shows top 10 global producers of natural gas in 2019. While the USA and Russia are top producers, they are also top consumers of gas, but Russia exports higher fraction of it.

We can also see from BP Statistics (2020) that in 2019 (the last year before covid2019 pandemic, resulting in some decline on energy consumption) the lion share of natural gas to EU has been delivered from Russia (170.6 bcm on pipeline and 20.5 bcm of LNG). The next supplier was Norway, with 109.1 bcm of pipeline and 5.8 bcm of LNG. The third supplier was Algeria with much smaller volumes:

21.4 and 9.5 bcm. The largest volume of LNG came from Qatar (29.7 bcm), followed by the USA (17.2 bcm) and Nigeria (13.3 bcm).

Figure 8.3 shows top 10 global consumers of natural gas in 2019. We see that EU is the second largest consumer of gas, but it is the largest importer of it.

The largest natural gas consumer in EU in 2019 was Germany (88.7 bcm), followed by UK (78.3) and Italy (70.8).

Energy scarcity in EU was observable already in 2021, when electricity prices were sky-rocketing in many EU countries. For example, wholesale electricity prices in Spain (normally below 60 Eu/MWh) grew to 200 Eu/MWh in October 2021 (before the war in Ukraine) and stayed above this level till May 2022 (Statista 2022). This problem caused a lot of internal problems in Spain and was widely discussed by Spanish TV. Note that this was an internal EU problem, not related to imports of energies.

Since February 24, 2022, the physical supply of natural gas to EU became much lower, due to sanctions over Russia (for the war with Ukraine) and later due to physical destruction of the North Stream, the main route of pipeline gas supply between Russia and EU.

The dynamics of natural gas prices for last 5 (and 25) years (in USD/MMBtu) can be tracked using site https://tradingeconomics.com/commodity/natural-gas. As we can see, between 2010 and 2020 prices have been close to 3 USD/MMBtu (and practically never above the level of 5, contrary to the period 2005–2008, when they were high and volatile between 7 and 14). In the second half of 2021, they rose from 3.5 and 5.5, while in March–December 2022 they were between 5 and 9.4. Since 2023 prices have returned to the levels below 5, but were still higher than in 2019–21. It is also possible to see the graph in Euro/MWh; the peak price of 340 Euro/MWh was reached in August 2022, while it stayed below 50 Euro/MWh before mid-2021 and after March 2023. This graph shows high sensitivity of market to gas shortages.

8.5.1 Politico-Economic Reasons for Gas Scarcity

The gas conflict between Russia and Ukraine in 2009 has shown European vulnerability to temporal interruption of gas supply, especially for Balkan region, with insufficient storage capacity and pipeline connection with other EU regions (Wirl & Yegorov, 2009). At present, it becomes higher.

Natural gas scarcity in EU today has politico-economic rather than physical reasons: sanctions to Russia because of the war in Ukraine and explosion on North Stream. EU has no gas pipelines with Middle East (where Iran can potentially replace Russia), but its LNG regasification capacities are not sufficient. What is more important is the limited global capacity of LNGs and competition for them with Asia.

Since 2008, natural gas prices became regional, being the lowest in the USA, the highest in Asia and of some intermediate level in EU. The reason is cheap production of shale gas in the USA and large and growing demand for natural gas in Asia, driven both by the economic growth in China and high demand in Japan after Fukushima disaster in 2011. Japan can consume the gas only in a form of LNG, while China partly consumes LNG and partly pipeline gas. Since EU demand for gas imports was not growing in the last decade, new LNG plants were oriented on Asian markets with steadily growing demand and high prices. That is why supply disruptions since 2022 created a very high increase of gas prices in EU.

Gas scarcity in Europe has geopolitical roots, because it is surrounded with many potential gas suppliers by pipeline, including Russia, Iran and Turkmenistan. In a free market environment, all of them could solve the problem of EU gas scarcity.

8.6 Conclusions

1. At present, we observe regional energy scarcity in Europe, which leads to negative economic consequences, including price rise for electricity and gas for consumers and producers, as well as high inflation, generated by this.
2. This chapter analyses the roots of this process. It is complex interaction between economic and physical processes and needs interdisciplinary approach. The overall situation in EU before 2020 was a quasi-stationary process of moderate transition to renewable energies, with more investment in solar and wind power stations, stabilization of natural gas consumption, decrease of coal and nuclear energy consumption in some countries.
3. A sustainable long-term plan for green transition should also include new capacities for HV lines in order to balance random wind supply in different regions, along with other energies. But EU had bottlenecks in cross-country transmission of energy, which did not allow for full integration of EU electricity markets in order to deal with this balance.
4. It is shown that for a proper balance of wind energy with itself, it is necessary to integrate regions of about 500–1000 km in diameter, and this includes integration of several EU countries in one market.
5. Supply security for natural gas depends also on the development of internal infrastructure in countries and between them. That includes pipelines and regasification plants. Contrary to electricity, gas can be stored in large volumes. Thus, interregional transmission is less important comparing to electricity. However, geopolitical factors are more important for gas. There was geopolitical shock in 2022, caused by sanctions over Russia, which has resulted in shortage of natural gas in EU in 2022 and sky-rocketing prices for it, causing energy poverty of the population, decline in production and welfare.

References

Archer, C., & Jacobson, M. (2007). Supplying baseload power and reducing transmission requirements by interconnecting wind farms. *Journal of Applied Meteorology and Climatology, 46*, 1701–1717.

BP Statistical Review of World Energy. (2020).

BP Statistical Review of World Energy. (2021).

BP Statistical Review of World Energy. (2022).

Eurostat. Statistics explained. https://ec.europa.eu/eurostat/statistics-explained/index.php?title=Renewable_energy_statistics#Share_of_renewable_energy_more_than_doubled_between_2004_and_2021

Haslett, J., & Raftery, A. E. (1989). Space-time modelling with long-memory dependence: Assessing Ireland's wind power resource. *Journal of the Royal Statistical Society. Series C (Applied Statistics), 38*(1), 1–50.

Marin, C. (2021, May). Gas pipelines and LNG carriers. Le Monde Diplomatique. https://mondediplo.com/maps/gas-pipelines

Opitz, S., & Tellman, U. (2015). Europe's materialism: Infrastructures and political space. https://limn.it/europes-materialism-infrastructures-and-political-space/

Statista. (2022). Spain, monthly wholesale electricity price. Statista.com. https://www.statista.com/statistics/1267552/spain-monthly-wholesale-electricity-price/

Tooraj, J. (2013). Security of the Europena electricity systems: Conceptualizing the assessment criteria and core indicators. https://www.researchgate.net/publication/281127145_Security_of_the_Europena_Electricity_Systems_Conceptualizing_the_Assessment_Criteria_and_Core_Indicators

Toshiaki, I. et al. (2009). Spatial correlation and temporal fluctuation of wind velocities observed on a plain field. 7th Asia-Pacific conference on wind engineering, Nov. 8–12, 2009, Taipei, Taiwan.

Treugoma.ru. http://treugoma.ru/electric-energy/feature-system/ (in Russian).

Wind Europe (2016) https://windeurope.org/wp-content/uploads/files/about-wind/statistics/WindEurope-Annual-Statistics-2016.pdf

Wirl, F., & Yegorov, Y. (2009). Ukrainian Gas Transit Game. *Zeitschrift fur Energiewirtschaft, 2*, 147–155.

Wirl, F., & Yegorov, Y. (2014). Renewable energy: Models, implications and prospects. In L. Bernard & W. Semmler (Eds.), *The Oxford handbook of the macroeconomics of global warming*. Oxford Univ. Press. ISBN-13: 978-0199856978.

World Energy Outlook. (2014). International Energy Agency.

Yegorov, Y. (2018). Natural gas infrastructure and energy security: Comparison between EU and China. In M. Radovanovic (Ed.), *Energy security: Perspectives, improvement strategies and challenges* (pp. 171–196). NOVA Science Publishers. ISBN: 978-1-53613-508-4.

Yegorov, Y., & Wirl, F. (2010). Gas transit, geopolitics and emergence of games with application to CIS countries. USAEE/IAEE WP 10-044, February 2010. https://papers.ssrn.com/sol3/papers.cfm?abstract_id=1560563

Yegorov, Y., & Wirl, F. (2018). The role of HV grid development for future security in electric energy. In M. Radovanovic (Ed.), *Energy security: Perspectives, improvement strategies and challenges* (pp. 71–96). NOVA Science Publishers. ISBN: 978-1-53613-508-4.

Chapter 9
Policy, Market, and Skills Barriers to Heat Pump Deployment in the United Kingdom

M. Cotton

Abstract UK domestic heat decarbonisation strategy is undergoing rapid policy change. The lag in energy efficiency performance improvements in domestic hot water and space heating—when compared to the electricity sector—is a key barrier to achieving national net zero policy targets. This chapter outlines the key issues raised in domestic space heating and hot water decarbonisation in the UK, with a specific focus on the uptake of air and ground-source heat pumps—a major component of government strategy for future homes. The key policy, market, and skills barriers to the rapid uptake of heat pumps are assessed, including issues of retrofitting older housing stock, encouraging market uptake of subsidised technologies, and retraining fossil-fuel intensive hot water and heating systems professionals to encourage consumer confidence. The key policy recommendations from this analysis are to reframe heat policy through a more holistic approach to both "supply push" and "demand pull" in the heat pump market—not only ensuring new building standards, but improving funding for skills, quality control and compliance in the sector to reduce costs, speed up market growth, improve consumer confidence, and ensure a "just transition" in the heating sector.

9.1 Introduction: Action on Low-Carbon Building Market Growth

Approximately half of global energy consumption is related to heat; accounting for 40% of global energy-related carbon emissions. In the UK, approximately 24 million end-users rely upon natural gas boilers for hot water and space heating. Transition away from this technology is an essential component of UK domestic decarbonisation and green growth strategy. The UK's commitments to cutting greenhouse gas emissions from buildings are linked to the contemporary social

M. Cotton (✉)
Teeside University, Middlesbrough, UK
e-mail: M.Cotton@tees.ac.uk

© The Author(s), under exclusive license to Springer Nature Switzerland AG 2024

T. Devezas et al. (eds.), *Global Energy Transition and Sustainable Development Challenges, Vol. 1*, World-Systems Evolution and Global Futures,
https://doi.org/10.1007/978-3-031-67583-6_9

and economic development policies described under the umbrella of the "Levelling Up" policy agenda. Within this policy platform, the concept of a "green industrial revolution" has gained political traction in the UK government's clean growth strategy. Former Prime Minister Boris Johnson outlined a 10-point plan for green industrial revolution, the tagline of which was to "build back better and build back greener". In the context of the built environment, this policy strategy is designed to stimulate improvements to residential, retail, and industrial building fabric efficiency, through insulation and other energy efficiency measures, making changes to the technologies used for both space heating and cooling within buildings, and improving the energy efficiency and performance of domestic energy services.

Notable among potential domestic decarbonisation measures within the Heat and Buildings Strategy is a commitment to provide 600,000 air/ground source heat pumps by 2028. Despite positive policy commitments to sectoral decarbonisation, in this chapter I explore the principal barriers to uptake of heat pumps within the new build and retrofit of domestic housing stock, with an emphasis upon policy, regulatory, market, and skills progress within the sector.

9.2 Buildings Strategy and Levelling Up Through Domestic Decarbonisation

The goals of net zero and socio-economic development through so-called levelling up policies, involve a range of market mechanisms designed to stimulate changes to energy systems within domestic and commercial properties. The Government's 2021 buildings strategy (BEIS, 2021) states:

> ...decarbonising buildings will help the economy grow, create new green jobs, and deliver greener, smarter, healthier homes and workplaces with lower bills. Delivering energy performance improvements and low-carbon heating systems will create new jobs in all parts of the UK – offering enormous potential to support our 'levelling up' agenda.

In practice, a net zero buildings strategy requires improvements to both the efficiency and design flexibility of domestic buildings (e.g. moving away from traditional/legacy housing stock designs through the Future Homes Standard discussed below), changes to UK supply chains, and support for alternative low-carbon heating technology options. At the household scale there are range of issues that influence the uptake of new low-carbon space heating technologies (I identify these summarised from the following: Hoggett et al., 2011; Singh et al., 2010; Snape et al., 2015):

- *Consumer knowledge and choice of heating technology* (including advice from installers).
- *Implementation factors* (such as "hassle costs" of installation, maintenance, and use).
- *Technology availability and supply chain management* for space heating in the context of growing national and global consumer demand for building materials

and components, and supply disruption from continued Covid-19 lockdowns in China and other global manufacturing centres.

- *Rapid cycles of innovation*: The rapid obsolescence of recent to-the-market technologies including early adoption of air and ground source heat pumps leads to consumer hesitancy.
- *Affordability of new heating technologies* given the upfront costs of installation under conditions of a national cost of living crisis, inflation, and rising interest rates affecting loan costs.
- *The relative market price of fuels*: Heat pumps are popular in some European nations due to the relatively low running cost. Heat pumps use a smaller amount of electricity than other electric heating methods that are normally used (e.g. electric fires or radiators). However, in the UK as of October 2022 the energy "price cap" puts the unit cost of gas at 10.3 p/kWh, vs. electricity at 34 p/kWh (Ofgem, 2022); thus the relative cost saving is lower than in some other European countries (e.g. France where electricity is €*0.1899* per kWh). This makes heat pumps comparatively less attractive to UK consumers.
- *Policy mechanisms to support market uptake*: Recent policy strategy stems from the Energy Security Bill. These include: the replacement of the Renewable Heat Incentive with a potential Low-Carbon Heat Scheme, The Home Upgrade Grant and Social Housing Decarbonisation Fund which provide targeted support to heat pump installations. Practical supply push measures include announced a 5-year zero VAT rating for heat pumps and their installation. This raises equity implications if government support for high-cost installation of low-carbon technologies that lead to energy savings are taken up by wealthier consumers (i.e. subsidy for wealthy consumers to install heat pumps upfront, that then lead to cost savings for them in the long-term does not benefit the poorest consumers).
- *Human resources* for manufacture, installation, modelling, project management, maintenance, and repair of novel technology solutions.

Against this backdrop of overlapping challenges, the Government's strategy in the low-carbon building sector is to (BEIS, 2021):

> [Support] 240,000 skilled, green jobs by 2035, concentrated on areas of the UK where investment is needed most.

The link between housing decarbonisation and economic opportunity is clearly articulated in contemporary UK policy frameworks. The UK has some of the oldest and most energy inefficient housing stock in Europe. Roughly 90% of existing homes are expected to still be in use in 2050, and so Government, social housing authorities, private sector developers, owners, and residents are tasked with collectively upgrading up to 27 million homes. The economic potential of a low-carbon heat transition is significant. At present total energy services are estimated at gross added value (GVA) of £26.5 billion to the UK economy (Brookes, 2021). Analysis of the business opportunities presented by low-carbon innovation in buildings, including domestic retrofit of existing building stock, are modelled at a peak of £5.7 billion gross value added (GVA) per annum and 82,000 jobs in the UK by the mid-2030s (Vivid Economics, 2019). The "primary" GVA potential of low-carbon buildings achieved through energy efficiency, growth in domestic markets for retrofit, design and construction of low-carbon homes, offices and manufacturing centres, and the export of designs, expertise and construction materials, and the sale of new electricity demand reduction and space heating technologies (including smart heating appliances) (Jones & Tam, 2023). The secondary economic benefits include lower electricity and gas bills for consumers (improving the spending power of households), and housing market uplift—as energy efficiency tends to improve property values (EPC C-rated properties are worth approximately 5% than EPC D-rated properties, all other factors being equal). The overall national-to-regional economic and social benefits of domestic low-carbon transition are clearly articulated across a range of academic and industry studies (Brown et al., 2021; Daruwala et al., 2022; Gaur et al., 2021; Hanna et al., 2016; Sovacool et al., 2020), making this an urgent policy priority to meet twin levelling up and net zero goals.

9.3 Heat Pump Installation

One key growth technology in the decarbonisation of domestic heat is the air or ground-source heat pump. The International Energy Agency (IEA) reports that around 190 million heat pump units were in operation in buildings worldwide in 2021, with the global stock increasingly steadily in primary heating markets—North America, Europe, and northern and eastern Asia despite global supply chain disruption in 2022 (Delmastro et al., 2022). In the UK, the *Building Services Research and Information Association (BSRIA)* estimated a 27% increase in average sales for heat pumps designed for commercial use in major European markets (BSRIA, 2022), driven primarily by concerns over domestic energy affordability and energy efficiency brought about by the reopening of the global economy leading after the easing of Covid-19 lockdown restrictions leading to a surge in fossil fuel demand, and the Russian invasion of Ukraine leading to economic sanctions and restricted gas flows to European markets (Pedersen et al., 2022), and the push for reversible heat pumps driven by twin space heating and cooling challenges in the face of climate change. However, against this backdrop of technology growth, heat pumps cover around 10% of the global heating demand in buildings, with deployment levels remaining too low to meet the Net Zero Emissions by 2050 Scenario. It is notable that 85% of heating systems operating in the UK are still natural gas boilers, and 1.5 million gas boilers are installed yearly, making it the biggest boiler market among European nations (SPRSUN, 2022). The IEA argue that further policy support and technical innovation to reduce upfront costs, remove market barriers to complex renovations, improve energy performance and durability, and exploit the potential of heat pumps as an enabler of power system integration and flexibility are needed (Delmastro et al., 2022).

Though growth in the heat pump sector is emerging, progress in the UK has stalled due to a lag in policy implementation. In 2013, the Climate Change

Committee advised the UK government that domestic space heating decarbonisation should aim to install 600,000 heat pumps *per year by 2020* to meet the 2050 net zero target. Seven years later (in 2020) this ambition was renewed as part of "The Ten Point Plan for a Green Industrial Revolution", with the government simply shifting the deadline back to install 600,000 heat pumps per year by 2028 (DESNZ 2020). Recent policy research shows that the UK government hadn't put any significant policy measures in place during that time to deliver on the recommendations of the CCC in 2013. As such, the heating sector will need to decarbonise eight homes every minute for the next 29 years to meet net zero by 2050 (Energy Catapult, 2019). Thus, although the potential economic and carbon savings opportunities from heat pump transition are promising, there remain a range of significant societal, policy and market barriers which stymie progress towards domestic Net Zero in the private ownership, rental, and social housing sectors. I divide these barriers into two broadly defined domains below:

1. *Policy and market barriers to heat pump installation.*
2. *Skills gaps in heat pump installation in low-carbon retrofit and new build properties.*

9.3.1 Policy and Market Barriers to Heat Pump Installation

Despite national commitments to low-carbon building construction and retrofit of existing domestic and commercial stock, multiple policy, regulatory and market barriers remain. I identify these barriers under the following categories:

1. Housing policy and buildings standards
2. Housing supply and new build properties
3. "Fabric first" retrofit of existing properties
4. Installation quality, consumer confidence, and trust

Housing Policy and Buildings Standards

The key national policy drivers to low-carbon domestic transition are the Heat and Buildings Strategy (HBS) (BEIS, 2021), the Future Homes Standard (Ministry of Housing, 2021) and the closure of the Domestic Renewable Heat Incentive and the implementation of its replacement: Low-Carbon Heat Scheme through the Energy Security Bill.

The proposed Low-Carbon Heat Scheme mirrors that of other industry-focussed policies designed for supply-push of low emissions vehicles or renewable electricity generation. The plans which will be supported by secondary legislation, will place obligations on manufacturers of fossil fuel heating appliances (e.g. natural gas boilers) to meet ever-rising standards for heat pumps sales as a proportion of their total appliance sales (through some combination of selling their own models or purchasing credits from other heat pump manufacturers—in a cap-and-trade-type approach). This provides a clear policy signal to the market, and simplifies the incentives for manufacturer heat-pump adoption, thus helping to build a heat pump supply chain and maximising consumer choice among manufacturers (Department of Business, 2022).

Industry incentives for supply push of heat pumps are matched by a regulatory framework from 2025, that makes it a core requirement for CO_2 emissions produced by new homes to be 75–80% lower than those built to current standards. The Future Homes Standard requires new build homes to be "zero carbon ready", with no subsequent retrofit work required to benefit from the decarbonisation of the electricity grid and the electrification of heating (Ministry of Housing, 2021). From 2021, the interim measure is that new homes are expected to produce 31% lower carbon emissions. Existing homes will also be subject to higher standards—with a significant improvement on the energy efficiency standards for extensions to improve fixed lighting, heating efficiency, and ventilation standards. The core elements of the Future Homes Standard, are that:

- New homes will not be built with fossil fuel heating (replacing the current standard of natural gas combi-boilers).
- Homes are intended to be future proofed with low-carbon heating and high levels of energy efficiency for lighting and in-built electrical appliances.
- No further energy efficiency retrofit work will be necessary to enable them to become zero-carbon as the electricity grid continues to decarbonise.
- Low-carbon heating systems will be integral to the specification of the Future Homes Standard and the Government anticipates that heat pumps and heat networks will become the primary heating technology for new homes.

Changes to building standards are one key policy mechanism to ensure technological change within the home. However, housebuilders are found to commonly aim for minimum compliance with efficiency standards, meeting only the most basic criteria at the lowest cost (Cullingworth, 2021; Pan & Garmston, 2012). Improved building standards codified in the Future Homes Standard may provide incentives to housebuilders to deploy existing innovations in low-carbon space heating such as air and ground source heat pumps but is unlikely to stimulate innovation or upscaling to go beyond existing performance standards. The regulatory environment for new buildings is therefore not a major driver of technological innovation to ensure future-proof net zero homes. However, regulatory change across the sector will likely encourage market uptake of heat pumps. Local authorities will play a key role in setting energy efficiency requirements for new homes that go beyond the minimum standards set through building regulations. Proposed new planning reforms will clarify the longer-term role of local planning authorities in determining local energy efficiency standards. It is vital therefore that local planning authorities aim for *ambitious* buildings efficiency standards across new build and extended homes across their respective regions.

Housing Supply and New Build Properties

Retrofit sits against a concern for overall housing supply within the economy. The 2019 General Election manifestos of all political parties included commitments to improve the housing supply in England. The Conservative manifesto pledged progress towards the target of 300,000 new homes per year by the mid-2020s, committing to at least one million more homes built over the next Parliament. However, greater clarity is sought over how this target of 300,000 units per year is to be achieved. There is a specific backlog of need amongst people living in unsuitable accommodation, and an affordability gap particularly for first time buyers, which prevents people from accessing the housing that they require, creating knock-on negative social and economic impacts such as overcrowding, impaired labour mobility, and increased levels of homelessness (Barton et al., 2022). The problems of housing supply were exacerbated during the Covid-19 pandemic due to slowdowns in house building, rising property prices due to a stamp duty holiday to encourage internal labour migration, and supply chain disruptions in building materials (Blakeley, 2021; Pfeifer & Steurer, 2022). The twin commitments of the 300,000 new homes per year building rate, and the low-carbon Future Homes Standard will put enormous pressure upon supply chains for heat pumps and associated wall, roof, and flooring insulation, and so government intervention through the low-carbon heat scheme is necessary to alleviate the cost "pinch points" within supply chains (including key materials, labour, and supply costs).

"Fabric First" Retrofit of Existing Properties

Changes to building standards only affect new build properties in the first instance. For the aging UK housing stock, a "fabric first" approach to low-carbon retrofit of existing buildings is the main driver of low-carbon transition. The logic of *fabric first* is to design processes that first maximise the performance of building components and materials (primarily in this instance: insulation—either wall cavity, loft or internal or external cladding), and then consider mechanical and electrical systems second (in this case heat pumps and solar PV and hot water systems). A fabric first approach improves electrical and thermal efficiency and reliability of the building itself, ultimately improving long-term cost efficiency (including reduced maintenance costs), and carbon emissions reduction. The fabric first approach counters the relatively high levels of embodied fossil fuel-based emissions involved in energy technology construction, and provides other benefits, such as passive efficiency as they do not require active user control (Oldfield, 2017). A "robust" fabric first approach is called for by multiple actors in the residential sector (Cotton et al., 2022; Pan & Garmston, 2012), yet this is highly dependent upon the level and type of skills, availability and supply chain efficiency across local housing markets. Moreover, there is a high "hassle" cost associated with fabric first retrofit, particularly when it involves changes to high use/high-cost room renovation such as bathrooms and kitchens, internal wall cladding (requiring the moving of furniture) or changes to electrical or hot water fixtures and fittings. The roll-out of domestic fabric first retrofit therefore operate around cycles of renovation that minimise disruption to homeowners/tenants.

There are essentially two models of retrofit—the one-off, whole-house approach, or a step-by-step approach over several years—the former is often construed as the least disruptive approach, though this is context specific, and neither the planned/emergent character of the retrofit nor the time scale over which they are completed seems to be linked to the carbon savings achieved, making a one-size-fits all approach impossible. Though a whole house approach reduced the overall timescale of disruption for the project, retrofit projects over an extended period of time allow residents to recover from the disruption of building works in between smaller projects, to save for the next phase of works, and plan retrofit in a way that fits with current investment patterns in housing (Fawcett, 2014). Whole-house retrofit would likely work best at the neighbourhood scale, in which a large volume of similar designed houses/flats all install a fabric first approach to insulation and then new low-carbon technologies. Policies adapted to a retrofit approach, commonly stress the economies of scale that come from this approach, particularly in relation to lower labour costs, and material purchases. However, this approach best suits social landlords who have the capacity for rolling retrofit across a large housing stock (Crilly et al., 2012), but is challenging in mixed-use developments where owner-occupiers, the private rented sector, and social landlords share buildings (such as flats) or adjoining properties.

Social housing providers are notable innovators in this space—they have control over the building management of large housing stocks, greater opportunity to access funding to retrofit on a large scale, and work to foster a close relationship between decarbonisation, resident quality of life and the place and communities they serve. They are therefore key "middle actors" who not only facilitate but also realise low-carbon transitions through retrofit (Cauvain & Karvonen, 2018), and provide key retrofit experience to contractors working in private owner occupier homes. The definition of net zero for the social housing sector is that housing associations need to eliminate a collective share of the 13% of direct, regulated carbon emissions produced by the burning of fossil fuels to heat space and water in residential buildings in all our current and future housing stock (Hughes, 2021). For most SHOs, the decarbonisation process is through roll-out retrofit, making use of the £3.8bn Social Housing Decarbonisation Fund promised in the Conservative Party manifesto (Grant, 2021). The roll-out retrofit approach has shown modest success for social landlords in the past. For example, Radian Housing, a social landlord with 16,000 properties, was funded by the European Regional Development agency for a series of 36 low-carbon renovations of near-identical houses in southern England. Gas and electricity consumption were monitored before and after the renovation works for a small number of these homes. The study showed that monitored gas consumption reduced by 50–60% whilst electricity consumption remained broadly unchanged or only showed small reductions during the same period (Killip et al., 2014; Radian, 2012).

The fabric first approach faces a number of market challenges. Davies and Osmani (2011) showed that the high capital costs for micro-generation technologies (such as solar-PV) and energy efficient materials; the disparity in VAT between new build and refurbishment; and the complexity of the UK existing housing stock are the most considerable low-carbon retrofit challenges for the residential sector. They recommend tax rebates on technologies and materials, removal of the VAT difference between new build and refurbishment, action to encourage more affordable micro-generation technologies (though notably solar-PV costs have come down considerably since 2011); and increased government supplied low-carbon programmes to drive successful retrofit.

The Government-led Heat and Building Strategy aims to ease consumer costs through a system of financial support mechanisms. These are primarily funded through the domestic Renewable Heat Incentive. However, from April 2022, the Clean Heat Grant provides support to households purchasing heat pumps. However, given the high upfront financial cost of installation, there is a risk that higher income households, motivated by pro-environmental action, are more likely to be early adopters and thus be more likely to take advantage of any government grants which are typically wound down over time (Branford & Roberts, 2022). As seen with previous incentives schemes (such as the feed-in-tariff), there is a risk that government taxpayer funded incentives for decarbonisation are taken up by the richest households. This is an issue of energy justice, in which the distribution of government-backed financial incentives disproportionately benefits with the wealthiest consumers. Ensuring mechanisms for uptake of Clean Heat Grants from low-income households (such as a staggered grant based upon means testing, with larger grants for lower income groups) would improve the overall fairness of the scheme.

Installation Quality, Consumer Confidence, and Trust

The considerable market barriers to retrofit affect consumer confidence and trust in the process and outcomes of low-carbon transition. There is growing consumer interest in domestic low-carbon transition, though this is not yet matched by market demand. A lack of impartial information on retrofit options, the support and finance needed, and long-term economic gains from energy saving are key barriers to uptake. Moreover, the limited consumer awareness or understanding about the benefits of improved building energy efficiency (specifically energy saving materials and technologies, and cost savings) often leads to insufficient incentives for interventions (Vivid Economics, 2019) and a narrative for energy efficiency that focusses primarily on minimising costs alone, rather than a fuller set of benefits related to saving costs alongside maximising health, improving comfort, and other benefits.

When it comes to disseminating information about the benefits of low-carbon heating systems, BEIS and the Heat Pump Association report that installers are the main contact for homeowners, and installers and tradespeople remain key trusted groups to advise on which heating system to install in the home (Heat Pump

Association, 2020). Making sure that installers can therefore confidently and accurately provide information on the range of social, health and cost benefits of a low-carbon system requires further training, assessment, and communication practices. However, even when consumers decide to proceed with installation work, they need to have confidence in the quality of the finished product, and that promised outcomes meet expectations. Research conducted through the Energy Catapult for The Gatsby Foundation found that the sector requires drastic improvements in quality and consistency across the residential sector. Too often, poor quality work either goes unnoticed, or is insufficiently penalised, not only creating disadvantage to consumers that receive expensive and poor-quality craft and materials, but also penalising those small and medium enterprises (SMEs) that perform good quality work by stimulating consumer distrust in the sector in the whole. Ensuring the quality of installation, maintenance and repair through active monitoring and assessment of installations would hold installers to account, therefore driving quality improvements across the supply chain (Energy Catapult, 2019). Also new retrofit buildings standards are aiming to improve the quality of installation. The new PAS2030 Installer Certification Standard, and 2021 MIS 3005 heat pump installation standard are now built into contemporary Building Regulations to ensure the quality of heat pump installations, and to raise standards in the delivery of retrofits by introducing additional skills requirements. Installers carrying out work as part of government schemes are now required to be registered with a licenced industry body, retrofit designs have to be carried out by an architectural technologist, architect or chartered surveyor (depending on the project), and all retrofits must be led by a trained project manager (NICEIC, 2022). However, there is currently no actual legislative requirement to be qualified or accredited to install a heat pump in the way that there is for natural gas boilers. Heat pump installers need to be Microgeneration Certification Scheme (MCS)-registered to receive payment for work done under government schemes such as the Renewable Heat Incentive or the incoming Boiler Upgrade Scheme, but there is no requirement for private sales to consumers. We argue therefore that the roll-out of MCS-certification requirements for all installers will help to establish consumer trust through timely, good quality and cost-efficient retrofit. This, in turn, will encourage rapid market uptake, which in-turn will grow the sector, providing further economic and social benefit gains.

9.3.2 Skills Barriers

Though increasing the market up-take (the "demand-pull" of low-carbon retrofit) is crucial, the housing industry still suffers a range of specific barriers to skills and human resource development in modern construction, buildings retrofit and the implementation of novel domestic space heating technologies (the "supply push" within the market). These collectively stymie the development of a cost-efficient and high-quality low-carbon transition service in the domestic residential sector. In terms of a skills gap, we identify two primary sets of challenges. The first concerns training in installation, maintenance, and repair of new innovations across the housing industry. The second is recruitment and retention of skilled staff. Review of the literature shows that overall concerns raised within the industry cover a range of issues within these two broader challenges (Bull & Domingues, 2021; King, 2010; Vivid Economics, 2019):

- The lack of training providers specialising in building physics, energy efficient design, technology design, maintenance, and repair.
- The porosity of skills markets and high demand for building sector services—i.e. highly trained individuals can take the skills elsewhere making it hard for SMEs to capture the full benefits of training.
- Limited opportunities for career progression for aspiring employees.
- High turnover of staff between employers, and to a lesser extend industry sectors.
- Limited requirements to demonstrate experience and accreditation for relevant skills.
- Limited ongoing training provision, access, and demand.
- Low levels of formal skills dissemination across the sector.
- Lack of partnership working across training providers.
- Limited support or funding for the SME sector to access training and skills.
- Limited demand for training amongst employees.

In the following section I break down these challenges into four component issues:

1. The size of the employment and training pool
2. Challenges of diversity and representation
3. Technology-specific installation challenges
4. Training provision challenges

The Size of the Employment and Training Pool

Though the UK heat pump industry will become a major player in the overall transition to net zero in the residential sector under the Future Homes Standard, at present the size of the market, the level of installation skills, inconsistent policy support and the lack of consumer demand has stimulated a major skills gap within the industry. The Government's Heat and Buildings Strategy (HBS) is key in setting out the overall vision for the transition to high-efficiency low-carbon buildings in the UK, but as shown by research conducted by the Social Market Foundation, the HBS is a "good start", but fails to address the extent of the workforce demands needed to make low-carbon domestic heat a reality (Norman, 2021). The HBS states that there are currently 1100 Microgeneration Certification Scheme (MCS)-registered and qualified heat pump installing *companies* in the UK (BEIS, 2021) (MCS is an accreditation scheme run by the Microgeneration Certification Scheme Service Company Limited), but provides no detail on the demographic make-up of the workforce (for example, how many are sole traders?), nor details any provision of tailored support for different types of suppliers.

The Heat Pump Association states that in 2019 there were an estimated 35,000 heat pumps installed in the United Kingdom and by 2020 the sector employed fewer than 2000 workers (Branford & Roberts, 2022; Heat Pump Association, 2020), though these estimates are difficult to verify (Norman, 2021). For comparison there are 130,000 registered gas boiler installers in the UK. Given the phase out of gas boiler installation in new homes from 2025, the current supply of skilled labour falls well below the numbers laid out in the 2011 UK Renewable Energy Roadmap (Department for Energy and Climate Change, 2011), which estimated that 150,000 jobs could be supported by the UK heat pump and biomass heat sector by 2020. The Heat Pump Association analysis of future market trends shows that the number of installers available in the sector needs to grow from this low base to around 70,000 by 2035 (Heat Pump Association, 2020). Using local data, in the Tees Valley in the Northeast of England, specifically, analysis using the EMSI Labour market analytics database report for all jobs within the 2021/22 "Engineering Activities and Related Technical Consultancy Industry Sector" found that of the 12,000+ jobs posted within this sector across the Northeast of England, 3% of these jobs were within the "Plumbing, heat and air-conditioning installation category". Though at the level of individual job postings, across the Tees Valley, we found only six unique postings for plumbing or boiler installation across 2021. It is unclear however whether the low rate of installer postings represents a low level of sectoral employment, or whether this represents uptake in the self-employed/SME sector. Searches for heat pump installers across the Tees Valley region turned up 11 results for small enterprises/sole traders, compared to 47 for gas boiler installation, representing considerable room for local growth within this sector. This is notable, given the Tees Valley's key role in the UK Government's levelling up policy agenda. If uptake is low within this key industrial region, then further examination of skills gaps nationally is warranted to meet national decarbonisation goals.

Challenges of Diversity and Representation

Branford and Roberts (2022) find that the impacts of a persistent heat pump skills gap reach beyond the jobs market and emissions reduction criteria—there are concerns about a *just transition* in the heat sector in which the social and economic benefits of heat transition can be felt across whole communities. They note that younger and older generations of workers are the ones most exposed to the risks of an industry skills gap. Younger people could potentially benefit from growing job opportunities in the heating sector, but low wages for apprenticeships, inconsistencies in quality of training and lack of routes into the sector need to be rapidly addressed. Older workers, including those moving from legacy carbon-intensive industries such as oil and gas or steel, are usually required to pay upfront to benefit from heat sector skills development, and often take unpaid leave to complete training. Lack of long-term government strategy on skills training in the sector have made personal investment in training a high-risk opportunity for many workers, thus exacerbating the skills gap challenge. Additionally, there is also an identified lack of diversity within the pool of trained employees, meaning that the heat sector (reflecting broader trends in the energy engineering sector) struggles to attract a diverse workforce. This not only reduces the talent pool available but also reduces the attractiveness of the sector overall, further limiting the future labour pool available. Despite attempts to address imbalances across the sector to attract workers across different backgrounds of gender, ethnicity, and disability, progress has been disappointingly slow at the sectoral level (Zekaria & Chitchyan, 2019). Collectively these skills gap barriers are cumulative. Thus, action to address pay, training costs, and active recruitment and retention of a workforce from a diverse array of demographics and social circumstances will not only grow the potential talent pool necessary to meet the 70,000-installer target by 2035 but will also address a "just" heat transition by alleviating problems of representation within the energy and engineering sectors (Bray et al., 2022).

Technology-Specific Installation Challenges

Trade Sector representatives interviewed by Branford and Roberts (2022) described the complexity of heat pump system installation. Unlike traditional gas boiler, heat pump installation requires a complex mix of skills including pipefitting, heat systems engineering, refrigeration, and electrical skills. Project installation is also more time-consuming than a traditional gas boiler, the external installation of the heat pump, electrical connections, new pipework, skirting boards, radiators, and a hot water tank is estimated to take roughly twice as long as a new natural gas-powered boiler system. Though typical installation takes an average of 2.5 days, complex projects can take "several weeks" if new electricity connection or fuse upgrade from the distribution network operator are required, or where radiators and pipework need to be replaced (Lowe, 2021). Also, though there are projected cost savings from heat pump installation, the process can cause unexpected risks, costs, and social impacts. For example, heat pumps can emit noise which breaches legal limits, and could perversely increase fuel bills if the right insulation is not installed first. Understanding the practical implications of heat pump installation under real-world scenarios for different housing types is a key research challenge. Of note in this regard is the Electrification of Heat Demonstration Project, led by The Energy Systems Catapult for the Department of Business, Energy, and Industrial Strategy (BEIS). This *£14.6 million demonstrator project will lead to the installation of heat pumps across 750 homes in the UK* across a representative range of housing archetypes and social groups (the majority on the gas grid). By monitoring the process and outcomes of installation and use, the demonstrator will provide useful data to assess the technical and practical feasibility of a large-scale roll-out of heat pumps (Energy Systems Catapult, 2021) and thus could potentially aid contemporary training in installation and project planning for different housing types and community contexts.

Training Provision Challenges

Of the 130,000 existing trained boiler installation practitioners, the Governments Heat and Buildings Strategy purports to upskill this workforce to install heat pumps through training that would take around 1 week to complete (BEIS, 2021). One of the core difficulties is providing sufficient foresight so that training providers can focus their efforts in a cost-effective manner, when trying to meet the skills needs of new and rapidly evolving markets, using new tech and installation practices (International Labour Organisation, 2011). As such, the Heat Pump Association (2020) report that contrary to the assertions of simple access to heat pump skills, the current route to becoming a heat pump installer remains costly, bureaucratic and confusing, with outdated content still being taught by many existing training providers. Low-quality or out-of-date skills training risks perpetuating a cycle of poor-quality installations as SMEs and sole traders pass on out-of-date knowledge to new apprentices and trainees within their organisations. This would ultimately damage consumer trust in the sector and further damage opportunities for market uptake of low-carbon heat systems.

Qualitative research with installers reports a varying degree of quality in training provision across the sector (Branford & Roberts, 2022; Norman, 2021). However, there is growing standardisation and certification emerging. The most-commonly recognised qualification is a what is called a Level-3 qualification in the UK (National Vocational Qualification (NVQ) or *Scottish Vocational Qualification* (SVQ)—Level 3 is immediately above the secondary education level) that complies with National Occupational Standards and is recognised by the Micro-generation Certification Scheme (MCS) registration bodies. This means that the NVQ/SVQ is commonly recognised as an acceptable qualification to join one of their respective schemes. This type of MCS-compliant heat pump training provision is provided by a range of education sector and private training suppliers. The NVQ/SVQ level *training typically covers how to* install, commission, handover, inspect, service, and maintain heat pump system installations non-refrigerant circuits (BPEC, 2022). Prerequisite learning for heat pump training is usually a minimum NVQ/SVQ Level 2/3 qualification in plumbing or conventional heating engineering, or a demonstrable number of years of relevant experience, plus demonstrated knowledge of Water Regulations/Byelaws. Acceptable prerequisite qualifications include NVQ/SVQ Level 2/3 in either Plumbing, Heating and Ventilating (Domestic Installation, or Industrial and Commercial Installation), Oil-Fired Technical Services, Gas Installation and Maintenance, or else Heating installation experience (minimum 3 years) in wet central heating system installation shown either by manufacturer courses certification or Gas Safe register, OFTEC, MCS, HETAS registration (Worcester Bosch, 2022). Yet as shown in regional low-carbon skills assessments (Bull & Domingues, 2021), training providers need to develop the necessary training and short courses to enable people to learn, share, and practice low carbon and green skills, with a specific need for clear progression opportunities from levels 2 to 7. This includes the development not only of accredited NVQ-SVQ options, but also a range of alternatives including T-Levels, HTQs and apprenticeships leading to high-quality installation outcomes suitable to meet the needs of mass roll-out retrofit and new build programmes to meet government-led decarbonisation pathways in the residential sector. Growth in the provision of flexible, ideally fully funded short courses will support lifelong learning across the different regions UK, and thus alleviate problems of persistent under-representation of workers from marginalised demographics and cultural backgrounds within the energy sector.

Specifically, courses for low-carbon heating solutions, including retrofit technologies and energy efficiency alongside low emission vehicles and infrastructure need to be implemented as an urgent action. Though individual FE and HE providers currently provide a range of courses in heating and ventilation, the curricula need to develop, specifically with individual courses to address specific skills gaps. Establishment of coherent longer-term strategy of partnership working across local authorities, partner colleges and private training providers is needed to develop suitable low-carbon skills provision, with dedicated support to SMEs to access these services to train their existing (small) workforces at low/no cost.

9.4 Conclusions and Policy Recommendations

UK heat decarbonisation lags far behind stronger action in the electricity sector. Infrastructural challenges around building stock, the high-carbon "lock-in" of natural gas-heated homes, and lack of policy leadership on decarbonisation in the past decade have slowed progress towards the deep decarbonisation action needed to meet net zero targets by the end of the decade. Making up for the lag in socio-technical design of efficient heating infrastructure and policy inaction means that rapid technology change is needed within the housing sector at an unprecedented scale. The challenges of post-covid restructuring of housing markets, and the looming potential of a "deep freeze" of housing markets in the UK followed a period of sustained house price growth followed by a rapid increase in interest rates resulting from global pressures on supply chains and UK government policy decisions (Stepek, 2022), means that further skills losses from the sector are likely. This creates a near-impossible challenge for the installation of heat pumps at the rate needed. Though housing policy is setting ever-higher standards for new build properties, concerns are raised around the capacity of supply chains and labour markets to meet the demand. What is needed is a more holistic approach to heat decarbonisation strategy across government strategy. Supply push of heat pumps requires work to build consumer confidence, this means providing good quality funded training for existing heating and gas engineers, alongside training on consumer advice and transparent provision of heat pump service costs, fabric upgrades and available subsidies. Regional mayoral authorities and local authorities must work with local training providers including colleges and universities to provide NVQ Level 2/3 qualifications at no/low cost to contractors to change the landscape of low-carbon skills across regions, and boost the rate of heat pump installation, and improve equality, diversity, and inclusion outcomes across the energy sector. The subsidisation of heat pump technologies should also follow a two-pronged approach, focussing not only on improving supply push to market, by targeting heat pump manufacturers and installers, but also "demand pull" using means-tested grants to target the users within the poorest communities, not those already capable of funding heat decarbonisation. By coordinating these policy actions, government renewable heat strategy can thus further the goals of a just transition, serving as a mechanism to further community renewal alongside decarbonisation, and thus achieve true "levelling up" within the sector.

Acknowledgements Research for this policy review chapter was conducted as part of the "Towards a Greener Tees Valley" project funded by the UK Government through the UK Community Renewal Fund. Thanks go to Professor Natasha Vall, Professor Paul Van Schaik, Thirteen Housing Group, and the Tees Valley Combined Authority for their assistance in this research.

References

Barton, C., Wilson, W., & Booth, L. (2022). *Tackling the under-supply of housing in England* (Issue). https://commonslibrary.parliament.uk/research-briefings/cbp-7671/

BEIS. (2021). *Heat and buildings strategy* (CP388). https://assets.publishing.service.gov.uk/government/uploads/system/uploads/attachment_data/file/1044598/6.7408_BEIS_Clean_Heat_Heat___Buildings_Strategy_Stage_2_v5_WEB.pdf

Blakeley, G. (2021). Financialization, real estate and COVID-19 in the UK. *Community Development Journal, 56*(1), 79–99.

BPEC. (2022). *Award in the installation and maintenance of heat pump systems (non-refrigerant circuits level 3)*. BPEC. https://bpec.org.uk/qualification/qcf-level-3-award-in-the-installation-and-maintenance-of-heat-pumps/

Branford, Z., & Roberts, J. (2022). *The installer skills gap in the UK heat pump sector and the impacts on a just transition to net-zero*. https://pure.strath.ac.uk/ws/portalfiles/portal/133553551/Branford_2022_Executive_summary_heat_pump_skills_gap_and_the_just_transition.pdf

Bray, R., Montero, A. M., & Ford, R. (2022). Skills deployment for a 'just'net zero energy transition. *Environmental Innovation and Societal Transitions, 42*, 395–410.

Brookes, S. (2021). *Non-financial business economy, UK and regional (Annual Business Survey): 2020 results. Non-financial business economy, UK: Sections A to S*. Non-financial business economy, UK: Sections A to S.

Brown, D., Jaccarini, C., Kumar, C., Madge, C., & Powell, E. (2021). *Financing Wales' housing decarbonisation*. New Economics Foundation.

BSRIA. (2022). *World Market Intelligence's annual study on heat pumps*.

Bull, R., & Domingues, A. R. (2021). *Levelling up and greening out: A just and holistic transition to low carbon jobs. An analysis of challenges and opportunities for the UK, the Midlands and Bolsover*. http://irep.ntu.ac.uk/id/eprint/45411/1/1501101_Bull.pdf

Cauvain, J., & Karvonen, A. (2018). Social housing providers as unlikely low-carbon innovators. *Energy and Buildings, 177*, 394–401.

Cotton, M., Stubbs, R., Lorrimer, S., Leighton, C., Mountain, A., Segovia, E., Van Schaik, P., Vall, N., Imani, E., & Knowles, M. (2022, July 6). Energy justice in the home: A qualitative technology assessment of domestic low carbon innovation within a social housing network. In *People, place and policy annual conference 2022: Breaking down barriers: Increasing inclusion across society, Sheffield.*

Crilly, M., Lemon, M., Wright, A. J., Cook, M. B., & Shaw, D. (2012). Retrofitting homes for energy efficiency: An integrated approach to innovation in the low-carbon overhaul of UK social housing. *Energy & Environment, 23*(6-7), 1027–1055.

Cullingworth, J. B. (2021). *Housing and local government: In England and Wales*. Routledge.

Daruwala, A., Workman, M., & Hardy, J. (2022). Identifying and unlocking the value from heat decarbonisation in the United Kingdom. *Energy Research & Social Science, 89*, 102672.

Davies, P., & Osmani, M. (2011). Low carbon housing refurbishment challenges and incentives: Architects' perspectives. *Building and Environment, 46*(8), 1691–1698.

Delmastro, C., Briens, F., Martinez-Gordon, R., & Husek, M. (2022). *Heat pumps: More effort needed*. https://www.iea.org/reports/heat-pumps

DESNZ. (2020). *The ten point plan for a green industrial revolution* (Department for Energy Security and Net Zero, London).

Department for Energy and Climate Change. (2011). *UK renewable energy roadmap*. https://assets.publishing.service.gov.uk/government/uploads/system/uploads/attachment_data/file/48128/2167-uk-renewable-energy-roadmap.pdf

Department of Business, E. a. I. S. (2022). *Guidance—Energy Security Bill factsheet: Low-carbon heat scheme*. https://www.gov.uk/government/publications/energy-security-bill-factsheets/energy-security-bill-factsheet-low-carbon-heat-scheme

Energy Catapult. (2019). *Foresighting skills for Net Zero Homes: A report for the Gatsby Charitable Foundation*.

Energy Systems Catapult. (2021). *Catapult to manage £14.6m electrification of heat demonstration*. Energy Systems Catapult. https://es.catapult.org.uk/news/catapult-to-manage-14-6m-electrification-of-heat-demonstration/

Fawcett, T. (2014). Exploring the time dimension of low carbon retrofit: Owner-occupied housing. *Building Research & Information, 42*(4), 477–488.

Gaur, A. S., Fitiwi, D. Z., & Curtis, J. (2021). Heat pumps and our low-carbon future: A comprehensive review. *Energy Research & Social Science, 71*, 101764.

Grant, J. (2021). *COP26 - A briefing for housing associations - 10 September 2021*. https://policycommons.net/artifacts/2641294/cop26/3664053/

Hanna, R., Gross, R., & Parrish, B. (2016). *Best practice in heat decarbonisation policy: a review of the international experience of policies to promote the uptake of low-carbon heat supply*. https://www.researchgate.net/profile/Richard-Hanna-2/publication/311809970_Best_practice_in_heat_decarbonisation_policy_a_review_of_the_international_experience_of_policies_to_promote_the_uptake_of_low-carbon_heat_supply/links/5888ab5daca27254af894b22/Best-practice-in-heat-decarbonisation-policy-a-review-of-the-international-experience-of-policies-to-promote-the-uptake-of-low-carbon-heat-supply.pdf

Heat Pump Association. (2020). *Building the installer base for net zero heating*. https://www.heatpumps.org.uk/wp-content/uploads/2020/06/Building-the-Installer-Base-for-Net-Zero-Heating_02.06.pdf

Hoggett, R., Ward, J., & Mitchell, C. (2011). Heat in homes: Customer choice on fuel and technologies. In *Study for scotia gas networks*. Energy Policy Group, University of Exeter.

Hughes, R. (2021). *Defining net zero for social housing-Decarbonisation task and finish group discussion paper*. https://policycommons.net/artifacts/2641463/defining-net-zero-for-social-housing/3664222/

International Labour Organisation. (2011). *Comparative analysis of methods of identification of skill needs on the labour market in transition to the low carbon economy*.

Jones, C. P., & Tam, S. (2023). *Low carbon and renewable energy economy, UK: 2021. Estimates of the size of the UK's Low Carbon and Renewable Energy Economy (LCREE), including turnover and employment*. https://www.ons.gov.uk/economy/environmentalaccounts/bulletins/finalestimates/2021

Killip, G., Fawcett, T., & Janda, K. B. (2014). Innovation in low-energy residential renovation: UK and France. *Proceedings of the Institution of Civil Engineers-Energy, 167*(3), 117–124.

King, D. (2010). The great zero-carbon skills gap. *Construction Research and Innovation, 1*(1), 24–29.

Lowe, T. (2021). *Major challenges in persuading homeowners to install heat pumps, government admits*. Building.co.uk. https://www.building.co.uk/news/major-challenges-in-persuading-homeowners-to-install-heat-pumps-government-admits/5112066.article

Ministry of Housing, C. a. L. G. (2021). *The Future Homes Standard: 2019 Consultation on changes to Part L (conservation of fuel and power) and Part F (ventilation) of the Building Regulations for new dwellings. Summary of responses received and Government response.*

NICEIC. (2022). *PAS 2030 installer scheme*. NICEIC. https://www.niceic.com/join-us/pas-2030-installer-scheme

Norman, A. (2021). *Did no one call the plumber?: Analysis of the Heat and Buildings Strategy's approach to skills and workers.* https://www.smf.co.uk/wp-content/uploads/2021/11/Did-no-one-call-the-plumber-Nov-2021.pdf

Ofgem. (2022). *Default tariff cap level: 1 October 2022 to 31 December 2022*. Ofgem. https://www.ofgem.gov.uk/publications/default-tariff-cap-level-1-october-2022-31-december-2022

Oldfield, P. (2017). A "fabric-first" approach to sustainable tall building design. *International Journal of High-Rise Buildings, 6*(2), 177–185.

Pan, W., & Garmston, H. (2012). Building regulations in energy efficiency: Compliance in England and Wales. *Energy Policy, 45*, 594–605.

Pedersen, T. T., Gøtske, E. K., Dvorak, A., Andresen, G. B., & Victoria, M. (2022). Long-term implications of reduced gas imports on the decarbonization of the European energy system. *Joule, 6*(7), 1566–1580.

Pfeifer, N., & Steurer, M. (2022). Early real estate indicators during the COVID-19 crisis. *Journal of Official Statistics, 38*(1), 319–351.

Radian. (2012). *Retrofit South East: Project summary report*. http://www.radian.co.uk/images/stories/case_studies/retrofit_south_east_final_summary_report_feb_2012.pdf

Singh, H., Muetze, A., & Eames, P. C. (2010). Factors influencing the uptake of heat pump technology by the UK domestic sector. *Renewable Energy, 35*(4), 873–878.

Snape, J. R., Boait, P. J., & Rylatt, R. (2015). Will domestic consumers take up the renewable heat incentive? An analysis of the barriers to heat pump adoption using agent-based modelling. *Energy Policy, 85*, 32–38.

Sovacool, B. K., Martiskainen, M., Hook, A., & Baker, L. (2020). Beyond cost and carbon: The multidimensional co-benefits of low carbon transitions in Europe. *Ecological Economics, 169*, 106529.

SPRSUN. (2022). *The Market Trend of Heat Pumps in the UK*. Retrieved 12/09/2022 from https://sprsunheatpump.com/the-market-trend-of-heat-pumps-in-the-uk.html

Stepek, J. (2022). *The UK housing market is at risk of a deep freeze* (Bloomberg UK, Issue). https://www.bloomberg.com/news/newsletters/2022-12-12/the-uk-housing-market-is-at-risk-of-a-deep-freeze

Vivid Economics. (2019). *Energy innovation needs assessment*. https://assets.publishing.service.gov.uk/government/uploads/system/uploads/attachment_data/file/845654/energy-innovation-needs-assessment-building-fabric.pdf

Worcester Bosch. (2022). *Training requirements for heat pumps*. Worcester Bosch. https://www.worcester-bosch.co.uk/professional/training-requirements-heat-pumps

Zekaria, Y., & Chitchyan, R. (2019). *Literature review of skill shortage assessment models. EnergyREV Project Report, WP 6.3 D1: EnergyREV EP/S031863/1*. https://www.researchgate.net/profile/Ruzanna-Chitchyan/publication/349883872_Literature_Review_of_Skill_Shortage_Assessment_Models/links/6046191a4585154e8c864ad7/Literature-Review-of-Skill-Shortage-Assessment-Models.pdf

Chapter 10
Paving the Way for Sustainable Development in Small Island Developing States: Insights from Cabo Verde's Energy Transitions

João Simões

Abstract This chapter examines the outlook for energy transitions in Cape Verde, a small island developing state located in the Atlantic Ocean, off the western coast of Africa. It explores diverse dimensions, including energy profiles, national policies, institutional frameworks, and emerging challenges and opportunities. The primary research question we pose is: what challenges does Cape Verde face in its energy transitions, and how can these energy transitions foster overall socioeconomic development? Additionally, the chapter discusses the role of international cooperation. It draws on various sources, including scientific and gray literature, and official national documents, such as development plans and reports. By highlighting the challenges of national transitions in the context of Cape Verde, this chapter contributes to a more comprehensive understanding of energy transitions and sustainable development in small island developing states.

10.1 Introduction

Small island developing states (SIDS) are a diverse group of countries that face unique environmental and developmental difficulties. The number of states classified as SIDS is determined by the classification criteria as there is no single definition for SIDS. The existing literature discusses the classification of SIDS, related criteria, and underlying assumptions (Alonso et al., 2014; Bass & Dalal-Clayton, 1995; Herbert, 2019; House, 2013). It uses predominantly geographical, socioeconomic, and environmental dimensions to identify developmental and environmental similarities and challenges.

As stated by the United Nations Office of the High Representative for the Least Developed Countries, Small Island Developing States and Landlocked Developing

J. Simões (✉)
City University of Macau, Macau, China
e-mail: joaosimoes@cityu.edu.mo

© The Author(s), under exclusive license to Springer Nature Switzerland AG 2024

T. Devezas et al. (eds.), *Global Energy Transition and Sustainable Development Challenges, Vol. 1*, World-Systems Evolution and Global Futures,
https://doi.org/10.1007/978-3-031-67583-6_10

Countries (UN-OHRLLS),[1] SIDS are a diverse group of 37[2] UN Member States and 20 Non-UN Members/Associate Members that face "unique social, economic and environmental vulnerabilities," mainly located in the Caribbean, the Pacific, Atlantic, and Indian Oceans, and the South China Sea. The 1992 *United Nations Conference on Environment and Development*, held in Rio de Janeiro, Brazil, described SIDS as follows (United Nations, 1993):

> Small island developing States, and islands supporting small communities are a special case both for environment and development. They are ecologically fragile and vulnerable. Their small size, limited resources, geographic dispersion and isolation from markets, place them at a disadvantage economically and prevent economies of scale. (...) They are considered extremely vulnerable to global warming and sealevel rise, with certain small low-lying islands facing the increasing threat of the loss of their entire national territories, Most tropical islands are also now experiencing the more immediate impacts of increasing frequency of cyclones, storms and hurricanes associated with climate change. These are causing major set-backs to their socio-economic development. Because small island development options are limited, there are special challenges to planning for and implementing sustainable development. Small island developing States will be constrained in meeting these challenges without the cooperation and assistance of the international community.

Since the establishment of UN-OHRLLS in 2001, it has supported, monitored, and further implemented the *Barbados Programme of Action* (BPoA), launched in 1994; the *Mauritius Strategy*, adopted in 2015; and the *SIDS Accelerated Modalities of Action* (SAMOA) *Pathway* (2014), to build synergies with other internationally agreed development goals, including the implementation of the 2030 *Agenda for Sustainable Development* and *The Paris Agreement*.

The recommendations and actions envisioned by the SAMOA Pathway are generic, representing a global framework of guidelines for a vast and heterogeneous set of countries and territories. This heterogeneity is defined by GDP per capita, available resources, demography, and geopolitics. It led the OECD (2018) to conclude that "the differences among SIDS point to the need for tailored development approaches across the group," yet due to "the common challenges they face, there is scope for mutual learning and exchange of experiences." The UN-OHRLLS recognizes this view "because island communities are at the frontlines of a changing world, they are spawning a wide range of solutions, some of which are addressing transnational development challenges."

It is essential to remember the results of the *SIDS Accelerated Modalities of Action (SAMOA) Pathway Report* adopted by the UN General Assembly Resolution on November 14, 2014, especially the outcomes referring to sustainable energy. After recognizing that "sustainable energy, including enhanced accessibility to modern energy services, energy efficiency and use of economically viable and

[1]The United Nations Office of the High Representative for the Least Developed Countries, Landlocked Developing Countries and Small Island Developing States (UN-OHRLLS) supports 90 vulnerable Members States including 46 Least Developed Countries (LDC), 32 Landlocked Developing Countries (LLDC), and 37 Small Island Developing States (SIDS). Some countries in these last two groups also belong to the LDC.

[2]In April 2023, Bahrain was removed, at its request, from the SIDS list.

environmentally sound technology, plays a critical role in enabling the sustainable development of small island developing States," the Resolution supports, among other things, the following measures (United Nations, n.d.):

(a) "To develop a strategy and targeted measures to promote energy efficiency and foster sustainable energy systems based on all energy sources, in particular renewable energy sources, in small island developing States, such as wind, sustainable biomass, solar, hydroelectric, biofuel and geothermal energy;
(b) To facilitate access to existing financing mechanisms to increase capital flows for the implementation of sustainable energy projects in small island developing States on renewable energy and energy efficiency;
(c) To support investment in initiatives by and for small island developing States, in particular the "SIDS DOCK" indicative project pipeline of renewable energy and energy efficiency and conservation projects, as well as in the areas of capacity-building and human resources development and public education and awareness;
(d) To promote international collaboration to ensure the access of small island developing States to energy by, inter alia, strengthening their integration with regional and international energy markets and increasing the use of locally available sources of energy in the energy mix, joint infrastructure development projects and investment in production and storage capacities, in accordance with national legislation."

The UN-OHRLLS emphasizes the critical role of *South-South* and triangular cooperation to overcome challenges and help SIDS accelerate the implementation of SDGs and the SAMOA Pathway. The body expects cooperation to "mobilize the human and financial resources, technology, innovation, knowledge, and expertise that SIDS now need." The UN-OHRLLS's report, *Good Practices in South-South and Triangular Cooperation for Sustainable Development in SIDS* (United Nations Office for South-South Cooperation, 2021), describes various models of international cooperation, including "South-South and triangular practices," to support SIDS in accelerating progress toward sustainable development (UNOSSC, 2021). However, the empirical evidence on the long-term effectiveness of the aforementioned cooperation mechanisms remains preliminary. In particular, it fails to demonstrate how SIDS can unlock the full potential of their energy transformation.

The International Renewable Energy Agency's IRENA (2019)'s report, *A New World—The Geopolitics of the Energy Transformation,* foresees the impact of the energy transformations on SIDS as follows:

> Small Island Developing States (SIDS) will benefit most if they adopt renewable energy sources rather than fossil fuels. The import of fossil fuels now amounts to 8% of their GDP. Many SIDS are also extremely vulnerable to the effects of climate change. SIDS possess ample renewable energy sources, and renewable technologies can meet most of their domestic energy needs. The shift would cut import bills, promote sustainable development, and increase their resilience. International cooperation to support SIDS' renewable energy ambitions is growing substantially, and 13 SIDS have established 60–100% renewable electricity targets.

Furthermore, the report *Accelerating SDG7 Achievement Policy Briefs in Support of the First SDG7 Review at the UN High-Level Political Forum 2018* (United Nations, 2018), discusses the status of Least developed countries (LDCs), Landlocked developing countries (LLDCs), and SIDS and their progress toward achieving SDG 7 (Ensure access to affordable, reliable, sustainable and modern energy). It shows that the population of these vulnerable countries where access to energy remains a significant challenge is around 1.1 billion people. Approximately half of the world's population who live without electricity are in LDCs. In 2016, the LDC population with access to electricity was 44.8%; in LLDCs, it was 53.1%, and in SIDS, it was 76.3%. The key takeaways of the report are:

- "Ending energy poverty in vulnerable countries and ensuring that no country or person is left behind has to become a priority for all stakeholders in order to achieve the 2030 Agenda.
- Each country's transition to a sustainable energy sector involves a unique mix of resource opportunities and challenges. National plans and policies should be designed for the particular needs and resources of each country, with a mix of grid, mini-grid and off-grid solutions.
- All vulnerable countries face inefficiencies in power utilities, which impact their operations and financial viability, and deter the private investments needed to improve generating capacity, and transmission and distributions systems."

Furthermore, the report identifies the following necessary actions:

- "To achieve the global goals on energy in LDCs, LLDCs and SIDS it will be essential to act fast to create enabling environments for private sector investment and to promote attractive project pipelines. This will require well-functioning institutions, and policy and regulatory reforms to help build credibility with investors and effectively scale up private investment, leveraging public resources for country-level implementation.
- Development Finance Institutions (DFIs) and development partners should increase the funding allocated to sustainable energy in LDCs, LLDCs and SIDS as this will have an impact across different sectors, including most of the SDGs, accelerating poverty eradication, structural transformation, and reducing vulnerability to fluctuating global energy prices.
- Moving a project from initial plan to bankable project requires significant time and human and capital resources (to prepare feasibility studies, environmental impact assessments, and permits). Vulnerable countries need more targeted support from their partners for project preparation to fast-track progress.
- Enhance integration of regional/cross border energy infrastructure and institutions to ensure economies of scale and lower the unit cost of energy generation.
- Create cross-sectoral linkages between sustainable energy and other development priorities (e.g., clean water, gender equality, improved education, access to healthcare, and climate change) to engage multi-stakeholder partnerships to support an energy transition, and to increase development finance flows that have the potential for higher impact and harmonized planning.

- Lack of maturity in energy access markets and underdeveloped financial markets in vulnerable countries mean that DFIs will have to play larger role in catalysing energy access investment in vulnerable countries.
- Ensure that the national energy policies take into consideration the energy demand profile of the poorest people, and ensure access to affordable energy. The positive development impacts of sustainable energy can be broadest by targeting the poorest people, who would normally not benefit from modern energy. The focus should not only be on promoting minimum access to households."

In response to the small island developing states (SIDS) request for assistance in realizing the goals outlined in the *SIDS Accelerated Modalities of Action* (SAMOA) *Pathway*, the *SIDS Lighthouses Initiative* (LHI) was introduced at the United Nations (UN) Climate Summit in 2014. The initiative's partners have agreed upon a new goal of 10 GW of installed renewable energy capacity by 2030. The Initiative "brings together partners, including SIDS, developed countries, regional and international organizations, development and multilateral agencies, private companies, research institutes and non-profit organizations that support the development and implementation of the SIDS national, regional and inter-regional and global sustainable energy strategies. IRENA is the coordinator and facilitator of the Initiative" (IRENA, 2022).

Recent years have highlighted two principal problems facing SIDS: the economic devastation caused by COVID-19 and the intensification of climate disasters and phenomena. The abrupt limits on fuel supply chains were devastating for import-reliant countries like SIDS. COVID-19 also made it more challenging for small island developing states to carry out their energy transition plans. After a decade of unprecedented expansion, renewable energy capacity and financing have decreased in recent years. Despite these difficulties, SIDS and development partners have proposed remedies and further steps to speed up energy transition and improve economic recovery (IRENA, 2022):

- Deploying renewables is the best remedy to address the ongoing climate and COVID-19 crises. This will strengthen resilience to future shocks and propel the economy forward.
- International public finance actors have a critical role to play in facilitating access to finance and leveraging opportunities for foreign direct investments, especially by eliminating cumbersome bureaucratic procedures.
- A mixture of measures and actions could be integrated to stimulate green recovery and meet SIDS' needs. These could include tailored climate financing packages to suit the needs of SIDS, support for low-carbon and climate-resilient investments, and better access to data to enable the studies and research needed to access financing.
- Debt relief or suspension measures, blended finance initiatives, and risk mitigation solutions are crucial but must be accompanied by transparency and accountability.

- Ensuring an integrated approach between SIDS and the international community, strengthening regional collaborations and partnerships, addressing institutional inequalities, and leadership will be essential.

Up to this point, we have identified several challenges to sustainable development in SIDS. These include: a high dependence on imported fossil fuels; access constraints to financing mechanisms; vulnerability to climate change; limited access to modern energy services; inefficiencies in local power utilities and grids; and institutional weaknesses, which encompass inadequate policy environments. Our literature review has also identified strategies to overcome these challenges, highlighting the need for international collaboration, building local capacities, and creating cross-sectoral linkages between sustainable energy and other development priorities.

This chapter examines the outlook for energy transitions in Cape Verde, officially the Republic of Cabo Verde, a small island developing state located in the Atlantic Ocean, off the western coast of Africa. It explores diverse dimensions, including energy profiles, national policies, institutional frameworks, and emerging challenges and opportunities.

10.2 Cape Verde Overview

Cape Verde is a volcanic archipelago on the northwest coast of Africa covering about 4000 square kilometers spread over ten islands with a population estimated by the World Bank of nearly 600,000 inhabitants. Its GNI per capita is around 3190 US$/year. Its economy depends on international trade (exports of goods and services represent only 25% of GDP, but imports reach 60% of GDP). The country has a consistent diaspora, with a negative migratory balance of about 7000 people in 2020. However, this emigration contributes to the inflow of personal remittances, representing 246 million USD in 2020. In the same year, the net flow of foreign direct investment was 104 million USD, to which 152 million USD of net foreign aid for development was added (World Bank, n.d.).

In the *Strategic Plan for Sustainable Development 2017–2021* [PEDS] (Governo de Cabo Verde, 2018),[3] the government already recognized that Cape Verde "prevails as a country with a high degree of economic vulnerability and low income, compared to other small island economies, especially the TOP 10 small island countries. The difference is significant when considering the HDI (human

[3]Meanwhile, the preparation of the "Cape Verde Ambitions 2030" strategy is underway, which "will pay special attention to the challenges that determine the continuation of the path towards sustainable development, namely the dynamic insertion of Cape Verde in the world economic system, sustainable tourism, economic diversification, decentralization, regional development and convergence, digital economy and innovation, human capital development and new partnerships for sustainable development". Validating the definition of this ambition is a "Declaration of Commitment to Sustainable Development" approved at the session of the National Assembly held in the city of Praia on September 29, 2020.

development index) variable and GDP per capita." Its vulnerabilities are critical in areas such as food (food insecurity is one of the main aspects of poverty in Cape Verde, particularly in rural areas) and international economic relations. Goods exports cover only 5% of the value of imports and are particularly important concerning strategic goods such as food and energy. Furthermore, there are pressures on the state budget derived from insularity and the rugged orography of its many islands. Moreover, Cape Verde has security concerns due to its location at the crossroads of intercontinental routes, making the country sensitive to threats from drug and human trafficking and other international crimes.

Environmentally, it is essential to note Cape Verde's aridity, which reduces its food production capacity, loss of biodiversity, forestal resistance difficulties, and accumulation of solid waste. However, as mentioned in the PEDS, "The development of a local market for renewable energies has a great potential to induce, directly and indirectly, the emergence of new businesses, industries, and services engaged in the construction, marketing, installation, and maintenance of renewable technologies, with a positive impact on the creation of new jobs, especially for young people, and on sustainable growth" (Governo de Cabo Verde, n.d.-a).

Cape Verde is an oceanic state with an Exclusive Economic Zone approximately 180 times larger than its land space, and "inshore fishing represents an important source of nutrition and income for population (....); this activity has been highly affected by overfishing and the impact of climate change, bringing several negative effects, including the loss of traditional fisheries, reduced income and jobs, and major concerns about food security" (PEDS). In addition to all possible mitigation for each vulnerability, it is essential to note the importance Cape Verde attributes to endogenous resources, education,[4] and integration into international development agendas such as *The 2030 Agenda for Sustainable Development, the SAMOA Pathway*, and *Agenda 2063*, and *The Africa We Want*. It is a multi-scale strategy seeking development in the context of global cooperation.

The strategic perspective of this country can be summarized with the statement of Prime Minister Correia e Silva, in September 2020, in the (UN News, 2020) that Cape Verde is "a country where it rains little and therefore needs a strategy to diversify the use of water, using desalination, and a strong link with renewable energies." Correia e Silva also highlighted the challenges and opportunities of climate change in the Community of Portuguese-Speaking Countries (CPLP), "First, we have a good alignment on fundamental issues, which have to do with the energy transition and the sustainable management of marine resources, because these are countries that have coastlines and a vast ocean surrounding them. Secondly, we are making the water strategy an important element at the center of concerns. Therefore, if fixed at the level of CPLP countries and Portuguese-Speaking

---

[4] In this case, as mentioned in the PEDS, Cape Verde's performance has been particularly positive and this is due to the strategic role assigned to education for development, which translates into, for example, a significant share of 20% of the budget in education expenditures and the impacts achieved on one of the highest literacy rates, or the population's education levels, on the African continent.

African countries, this whole strategy can give us more capacity for intervention and joint action." These declarations systematize and integrate the country's critical challenges (water scarcity, the use of its resources, and energy transition) within a management framework that promotes sustainability, looking to the future, and joint actions with its partners.

10.3 Strategic and Development Plans

This analysis of the Cape Verde situation, particularly the existing challenges, maintains that the pandemic crisis and war in Ukraine have prolonged and accentuated some development paths that the *Strategic Plan for Sustainable Development* (PEDS II) seeks to align with its four fundamental pillars: economy, social, environment, and sovereignty. This 4-year plan is supported by the *Master Plan of the Electric Sector 2018–2040*, "a structural document for the development of the electricity system, considering the a structural document for the development of the Electricity System, considering the main areas of development of the sector: territorial consumption forecasting, new investments and reinforcements in transport and distribution infrastructure of electricity, producer structure (location of plants, size, energy sources and technologies) and network management, institutional and organizational structure. This Plan covers the following topics: Electrical demand forecast; Needed offer to meet the electrical needs; Transport network expansion plan; Network analysis of the electrical system and least-cost optimization (lowest-cost); Economic and financial analysis; Investments and indicative costs in generation and transportation, impacts on electric energy tariffs; Environmental and social impact assessment. Environmental, economic and financial impact assessments, tariff studies as well as an Action and Investment Plan, together with Energy Policy Recommendations, complemented the evaluation of the strategy" (Governo de Cabo Verde, 2019).

In the context of PEDS II, the "*Energia 2023*" operational program was established based on the assessment that "Cape Verde plans to undertake a major shift towards the low-carbon economy by increasing the share of renewable energy sources in the electricity supply from 18% to 30% in 2025 and up to 50% in 2030. With adequate support, Cape Verde has indicated that the target for the share of renewable energy sources may go up to 100% by 2040. Cape Verde also plans to secure on-grid or off-grid electricity supply across nine islands and reach 100% access to electricity for all consumers by 2026" (Governo de Cabo Verde, 2023).

The second edition of the National Program for Energy Sustainability (PNSE II) maintains that the long-term goal signals the need to accelerate the transition to a secure, efficient, and sustainable energy sector without dependence on fossil fuels while guaranteeing universal access and energy security.

As seen above, energy sustainability in Cape Verde will be far from assured because the country depends on imported fossil fuels to satisfy its energy needs. However, the *Strategic Plan* recognizes that the country has a "significant potential for renewable resources" and there is a "high level of losses in the distribution and commercialization of electricity, poor operational performance of the concessionary companies and high energy prices, causing a substantial burden on company and family expenses."

Despite the path taken, this is a recognition of what still needs to be done during the energy transition for the country to comply with *SDG 7, Accessible and Clean Energy*, which represents Cape Verde's *Nationally Determined Contribution* (NDC). Despite the positive performance, renewable energies only represent 22% of the global supply of primary energy sources (TPES). It should be noted that the most significant contribution of renewable sources is wind energy (onshore) and, over time, solar energy. Other renewable energy sources are not part of the supply in Cape Verde, despite its insular nature and the opportunities of offshore space. Finally, the contribution that off-grid sources are making to renewable energy, which, although still insignificant (0.846 MW in 2020), has experienced a significant increase since its beginning in 2013 (0.074 MW).

To achieve these goals, "*Energia 2023*" proposes, among other measures, advancing the implementation of Energy Storage Systems (ESS) on all islands, promoting RE microgeneration (distributed generation) for self-consumption, and the "introduction of demonstration projects, namely: energy efficiency in public lighting (street light replacement with LED bulbs), promotion of energy efficiency in appliances and buildings (replacement of incandescent and fluorescent bulbs), energy efficiency in public administration, sustainable schools project, energy efficiency in industry and tourism, mass introduction of solar thermal systems, incremental growth in access to clean cooking technology (changing from firewood to liquefied petroleum gas (LPG), e-mobility, fostering scientific and technological research into wind, solar, geothermal energy and ocean thermal energy, and green hydrogen" (Governo de Cabo Verde, 2023).

"*Energia 2023*" considers that there are strong reasons to invest because the measures and projects referred to above allow increasing access to green and modern electricity, reducing "energy costs for households, industries public buildings and the cost of water for agriculture," reducing greenhouse gas (GHG) emissions, reducing fossil fuels imports, increasing renewable energy, reducing electrical grid losses, reducing external dependence on fossil fuels and "making the economy more competitive by reducing fixed energy costs" (Governo de Cabo Verde, 2023).

Energia 2023s optimism is supported by experts who have studied the Cape Verde renewable energy situation, including Erik Nordam, a research professor at Grand Valley State University. Unlike most Sub-Saharan African countries, in Cape Verde, most of the population has access to electricity, and consumption per person is well above the regional average. The electricity network is still not exempt from the endemic weaknesses of diesel generators, and power outs remain frequent. Cape Verde has considerable renewable resources: "The country has wind resources equivalent to those of Morocco, the solar potential of the Sahel, geothermal resources close to those of Kenya, and marine energy comparable to that of many coastal countries," explains Erik Nordman. He specifies (Les Smart Grids, 2018):

Wind power was the first of these resources to be exploited. The trade winds of the northeast of Cape Verde are considered excellent for the production of wind energy.

Ana Monteiro, Cabeolica engineer, explains that increasing the share of renewables while maintaining the network balance is the equation that Cape Verde is trying to solve. The winds are strong and regular, often exceeding 9.0 m/s at 50 m from the ground (wind power is estimated to be profitable from 6.4 m/s). Wind electricity already provides 25% of the consumption of the archipelago's three main islands. Power cuts are less frequent, but the intermittence of the wind requires increased vigilance. To increase energy security, Cape Verde will invest in storage solutions. Lithium-Ion battery units will be delivered with the following wind or solar farms. Smart grids will be associated with the largest of these parks. The country is also testing an original solution, combining pumped storage stations with seawater desalination. This solution would make it possible to produce hydraulic electricity and pure water. Another experiment in progress is a study aimed at harnessing seawater thermal energy south of the island of Santiago. Finally, the archipelago's isolated areas will have photovoltaic microgrids and storage units. The plan is ambitious and will require significant investment for a small country. However, the certainty that the state will recover quickly should help Cape Verde find the necessary international funding (Les Smart Grids, 2018).

Similarly, the Government Program 2021–2026 (Governo de Cabo Verde, n.d.-b) presents its vision to accelerate the energy transition as one of the pillars of the axis in developing the green economy. In this context, the government defines its energy policy as focusing on reducing fossil fuel dependence and energy bills to ensure sustainable, modern, and affordable energy access. The goal is to reach 30% of electricity production from renewable sources by 2025, exceed 50% in 2030, reach 100% in 2040, and reduce energy intensity by 10% by 2026 by implementing the *National Energy Sustainability Program.* The government also undertakes to implement policies and measures to accelerate electric mobility by implementing a program to increase the national fleet of electric vehicles for the public administration and government members to at least 30% of all vehicles. It will also strengthen measures to increase electric vehicle charging infrastructure installation and foster R&D projects for vehicle-to-grid (V2G) infrastructure.

10.4 The Regional Context and International Cooperation

Regionally, Cape Verde is an essential bridge between Western African countries and the European Union. Besides being a member of the *Economic Community of West Africa States* (ECOWAS),[5] Cape Verde signed a *Special Partnership Agreement* with the EU in 2007, making the Cape Verdean government a privileged actor

[5]The ECOWAS Center for Renewable Energy and Energy Efficiency (ECREEE) was established in Cape Verde in 2009 with the support of the ECOWAS Commission and the governments of Austria

in EU-ECOWAS relations. This *Special Partnership* permits Cape Verde to access EU funds to finance development projects. Such is the case of the Cape Verdean "*Centre of Renewable Energies and Industrial Maintenance*" (CERMI) (ECREEE, 2015) inaugurated in 2015. Financed by the EU, CERMI has attracted the partnership of several African countries, such as Burkina Faso, Senegal, Mali, Niger, Ghana, Nigeria, and the ECOWAS *Center for Renewable Energy and Energy Efficiency* (ECREEE), headquartered in the Cape Verde capital, Praia (ECREEE, n.d.-b). Therefore, Cape Verde enjoys international cooperation mechanisms and is an international cooperation entity advancing energy transition in Western Africa because ECREEE is a hub for African SIDS. Furthermore, Cape Verde is part of *The Power Africa Program* created in 2018 by the *United States Agency for International Development* (USAID) and ECREEE.

Despite being a full member of ECOWAS (a regional organization for energy access concerns and energy sector common policy), Cape Verde's insular reality separates it from the Western African regional energy market. In 2019, under the scope of the operationalization of a regional electricity market in the ECOWAS region, the West Africa Energy Group (WAPP) started to implement a policy plan for the development of regional infrastructure for the production and transmission of (renewable) energy for 2019–2033. ECOWAS launched a three-layer renewable energy production and distribution strategy (hydrographic, solar, and wind), but Cape Verde was omitted. However, ECOWAS has invested in Cape Verde's renewable energy expertise through ECREEE headquarters (ECREEE, 2020).

CERMI, besides being an educational institution with the principal objective of disseminating knowledge and developing skills in renewable energy and industrial maintenance, also advises on and assesses renewable energy for the Cape Verde Government.

It is essential to highlight the importance CERMI plays as an organization that attracts and relates to other institutions linked to renewable energies, including ECREEE, the Cape Verdean Government, and ALER, *The Lusophone Association for Renewable Energy*, to develop partnerships and prepare a joint agenda. This cooperation strategy is a step forward for Cape Verde and the African continent in their connection to European organizations. Following the example of cohesion and integration in the European Union (e.g., EU Directives with binding renewable energy targets), the ECOWAS Commission has gradually addressed climate change and renewable energy. The creation of ECREEE in Praia, Cape Verde, in 2010 was a significant step in bringing together Western African countries' plans and policies on this critical matter. ECREEE's action is aligned with the goals of ECOWAS 2020: (1) "A region that anchors its development on sustainable development, including agricultural and mineral resource development strategy, planned agricultural and industrial strategies, developing infrastructure and making services accessible to its citizens and enterprises." (2) "A region that conserves its environment and resources,

---

and Spain, along with technical assistance from the United Nations Industrial Development Organization (UNIDO).

promotes equitable and sustainable development in economic, social and environmental fields and brings its contribution to the resolution of the common problems and challenges confronting the planet" (ECREEE, n.d.-b).

International cooperation allowed Cape Verde to participate in ECOWAS' Thermic Solar program (with Nigeria, Burkina Faso, Ghana, and Senegal) and the ECOWAS effort to increase solar energy in Western Africa. ECREEE coordinates the program implemented by AEE INTEC. The SOLTRAIN program is a technical instruction initiative considered vital in implementing renewable energy in the region (Governo da Guiné-Bissau, n.d.).

The *ECOWAS Regional Centre for Renewable Energy and Energy Efficiency* (ECREEE) coordinates and executes programs, projects, and activities, including (ECREEE, n.d.-a):

1. Creating and efficiently managing an effective regional sustainable energy promotion agency.
2. Creating and implementing tailored sustainable energy policy and regulatory support frameworks.
3. Strengthening and applying various sustainable energy aspects.
4. Strengthening knowledge management, awareness raising, advocacy, and networks.
5. Mobilizing and implementing investment in sustainable energy projects and businesses.

It is also important to highlight Cape Verde's presence in an association that brings together 12 energy regulators from 6 Portuguese-speaking countries, RELOP—Association of Energy Regulators of Portuguese Speaking Countries. RELOP aims to promote cooperation and sharing of experiences, inspiration, and innovation to guarantee the protection of present and future consumers through a robust regulatory policy, integrating the entire Portuguese-speaking community based on the Portuguese language (RELOP, n.d.-b). Members of RELOP face common challenges in their energy transitions, including: attracting investment; increasing the share of renewable energy in the energy matrix; monitoring the implementation of public policies in the energy sector for a fair and inclusive transition; security and diversification of energy sources. RELOP's Strategic Plan 2022/2025 includes these challenges as courses of action to be dealt with collectively (RELOP, n.d.-a).

The *African Development Bank* (AfDB), established in Cape Verde in 1977, has assisted the nation in numerous projects, including the economy, agriculture, industry, water and sanitation, energy, transport, and governance.

The Renewable Energy and Improved Utility Performance Project (REIUP) for Cape Verde received World Bank approval in December 2021 for a 3.5 million USD International Development Association credit and a 3.5 million USD International Bank for Reconstruction and Development loan. The Global Infrastructure Facility will provide a 2 million USD grant, and the Canada Clean Energy and Forest Climate Facility (CCEFCF) will provide 7.5 million USD (of which 0.5 million USD is a grant and 7 million USD is a loan) to the project (World Bank, 2021).

Cape Verde has attracted the attention of major world powers to support their quest for cleaner and sustainable energies. In 2015, the Chinese Ambassador Du Xiaocong met Cape Verdean President José Fonseca and discussed Cape Verdean expectations concerning Chinese support in three areas: economy, tourism, and renewable energy (Expresso das Ilhas, 2019). In 2018, at the *China-Africa Cooperation Forum*, Cape Verde and China signed a memorandum of understanding for cooperation for *China's Belt and Road Initiative*.

The USA Department of Commerce, through its International Trade Administration, recently classified Cape Verde's renewable energy sector as a good business opportunity, stating, "Cape Verde offers tax benefits for investments in the sector and has a master plan for the electricity sector that identifies sites for the development of renewable energy projects. It is the government's stated intention to privatize ELECTRA, the water and energy production company, in 2022" (U.-S. Department of Commerce, 2022).

The *Special Partnership* between the EU and Cape Verde will bring a 24 million Euro investment to Cape Verde during 2021–2024. One of these investments' priorities focuses on developing green energy sources and green investment (European Commission, n.d.).

The Portugal—Cape Verde Strategic Cooperation Program 2022–2026, signed on March 7, 2022, defines six significant areas of cooperation between the two countries: (1) Education, Science, Sports, and Culture; (2) Health, Social Affairs, and Work; (3) Justice, Security, and Defense; (4) Environment, Energy, Agriculture, and Sea; (5) Public Finance, Economy, Digital, and Infrastructure; (6) Cross-cutting Areas (Instituto Camões, n.d.).

Other European countries have joined Cape Verde in developing energy sector investment projects. Co-financed by the European Union and Spain, the project "*El Pacto de los Alcaldes para el África Subsahariana*" promotes energy efficiency in the Municipal Council of Ribeira Grande de Santiago area by installing solar panels. The Cidade Velha local government will be able to make energy production 100% sustainable at its headquarters and the city's heritage points through the NAMA Facility. This European fund supports Cape Verde's government strategy that established the goal of replacing the country's entire vehicle fleet with electric vehicles by 2050. Locally generated renewable energy will replace imported fossil fuels in road transport. Thus, the government aims to relieve trade balance pressure while reducing air pollution, noise, and greenhouse gas emissions (Cooperación Española, 2022).

Angola selected Cape Verde to start its investment internationalization. The enterprise Anglo-Bal, among other areas of engagement, operates renewable energies, particularly photovoltaic ones (Câmara Comércio Sotavento, n.d.).

The United Nations supports Cape Verde's clean energy projects through various agencies and programs, including the UNIDO-sponsored Porto Novo photovoltaic project. The installed solar power plant has led to a decrease in operating costs at the Porto Novo Power Station. This reduction has impacted the cost of producing drinking water, which has decreased by approximately 5%. As a result, about 10,000 residents of the city of Porto Novo now have wider and more equal access to drinking water. This change has positively impacted their health and well-being (ALER, 2022).

Since 2006, Cape Verde has been successful in building a network of large and small dams, large desalination plants, and small projects to provide farmers and rural communities with access to clean, affordable water through the desalination of brackish water while utilizing only (or mostly) solar energy. In 2019, Cape Verde attracted a European Union-sponsored program focused to exploring wave electric energy generation. The study has identified Maio Island as an ideal location for wave energy generation, which can meet roughly 20% of the energy needs of the larger islands and most of the needs of the smaller islands. Cape Verde is developing into a hub of expertise in the fields of saltwater desalination and alternative energy, two topics that will be of the utmost significance to the entire world in the very near future (Leandro et al., 2023).

10.5 Institutional Frameworks for Energy Transitions and Sustainable Development

Cape Verde is part of international cooperation networks such as the UN and regional networks such as ECOWAS, the economic community of West Africa. It is also part of organizations with different development objectives, such as the CPLP. It is also allied with more focused objectives, such as those of the West African Power Pool (WAPP). WAPP aims to integrate the national power system and form a regional electricity market to provide regular and reliable energy at a competitive price to the ECOWAS region in the medium and long term.

These and other organizations that include Cape Verde establish a multifaceted cooperation framework to solve the country's challenges. Cape Verde also has objectives, guidelines, rules, partners, and investment sources essential for daily decision-making. These include the *Social Development Plan* and the *Government Program Master Plan for the Electric Sector 2018–2040.*

Concerning the involvement of the public and private sectors in energy transition, it is the government's task to define policies and manage resources. Therefore, it must equip itself with an operational organizational structure. Table 10.1 shows the present organization.

Other relevant actors, not considered in the above table, influence the country's energy administration. Teaching and training establishments should have their own research centers or partnerships that participate in research and development. Cape Verde has six universities with more than 200 approved degree courses. These include courses directly related to energy, including electrical engineering. The government must add professional training in energy transition-related matters to the country's higher education programs.

Table 10.1 Organization of the electricity sector in Cape Verde

Entity	Role
Ministry of Industry, Commerce and Energy—MICE Ministério da Indústria Comércio e Energia	• Defines the goals, strategies, and policies for the energy sector • Global supervision of the energy sector
National Department for Industry Commerce and Energy—DNICE—Energy Service—SE (Direcção Nacional da Indústria, Comércio e Energia de Cape Verde—Serviço de Energia)	• Plans and implements energy policy and strategies • Inspects ongoing projects and issues authorizations and operators' certificates • Stimulates programs for expertise enhancement • Promotes universal access to electricity • Proposes legislation and analyses the sector
Electricity production and distribution Public Enterprise—ELECTRA)	• The primary operator in eight of the nine inhabited islands
Electricity production and distribution Public Enterprise—Águas e Energia de Boavista (AEB)	• The primary operator on Boa Vista island
Economy Multisector Regulation Agency—ARME (Agência Reguladora Multissetorial da Economia)	• Regulates the prices and zeal for costumer protection
Cape Verde has six licensed energy operators	• Eólica, Cabeólica, ElectricWind, Solar, Electra, and Águas de Ponta Preta (APP)
Independent Power Producers/Consumers (IPPC)	• There are currently several IPPCs in Cape Verde. They are increasing considerably, but there is no data

Source: https://www.portalenergia.cv/

10.6 Discussion

Like many countries, Cape Verde has important challenges associated with climate change, contributing to mitigating the adverse effects and changing its infrastructure to ensure sustainability. In this sense, Cape Verde has positive characteristics of which cooperation with other countries in international, regional, or global organizations stands out. Furthermore, its stated desire to implement renewable energies instead of fossil fuels and the benefits attributed to investment in renewable energies helps create an overall picture of Cape Verde as firmly committed to the planet. This benchmark is further reinforced by Cape Verde hosting technical bodies such as ECREEE, equipped with solar and wind energy sources and significant experience in these fields. There is also strong evidence for potential wave and geothermal energy production. Additionally, valuing the circular economy and promoting efforts to end global warming effects can catalyze other industries, such as tourism, agriculture, and the economy of the sea. Global technological developments in renewable energy production, distribution, and consumption can also reduce energy costs and solve the problem of water scarcity through desalination.

These strengths are insufficient to overcome the weaknesses resulting from Cape Verde's location or the dispersion of its islands and have not yet made it possible to overcome the consequences of strong dependence on external financing and other international actors, the electricity grid losses or the intermittent nature of solar and wind energy production. Nor will they be enough to overcome the dramatic effects of climate change, which the country does not control. As an insular country, Cape Verde is particularly exposed to the adverse effects of global warming in the Atlantic Ocean latitudes of the Sahel. The country suffers from droughts, rising ocean waters, and destruction of marine biodiversity. Therefore, it embraced a serious effort to become autonomous in renewable energies (100% by 2050), aiming to contribute to international efforts concerning global warming and become autonomous in generating its domestic energy needs. Furthermore, the archipelago has embarked on a vigorous policy in favor of renewable energies by legislating to reduce fossil fuel use (blocking the import of internal combustion vehicles) and building schools focused on renewable energy technology.

Despite this balance, Cape Verde can benefit from the opportunities the global environment offers. It values its circular economy, technical, financial, and human resources availability, and the international community's focus on mitigating climate change. Cape Verde has a political commitment to achieve 50% renewable energy usage by 2030, is a member of all relevant global and regional organizations, and has a special partnership with the European Union. It also hosts the ECREEE. Additionally, Cape Verde has implemented policies and tax benefits to attract investment in renewable energy. Ongoing energy sector restructuring policy focuses on sustainable development and replacing fossil fuels (e.g., the privatization of ELECTRA, the Cape Verde water and energy production company, mini-grids for renewable energies to solve the existing electricity grid efficiency problem, tax benefits for investments in renewable energy, a political intention to stop importing fossil fuel powered vehicles beyond 2030), the ability to attract investment and international business, and a potential transformation into a specialist in renewable technology (such as wave power). Concerning security, Cape Verde is a stable country. Its waters do not suffer from the same problems as countries in the Gulf of Guinea. These strengths provide an optimistic outlook for the country's energy transition.

There are also threats to Cape Verde's energy transition. Global warming could have dramatic effects on the oceanic Sahel latitudes. Population overgrowth and climate and economic migrants could further stress the energy system.

10.7 Conclusion

This chapter examined the outlook for energy transitions in Cape Verde, a small island developing state located in the Atlantic Ocean, off the western coast of Africa. Cape Verde needs to mobilize significant financial resources over the next few years to invest in energy transitions, resources that could far exceed the savings made by decreasing oil/gas imports over the years. This gap will need to be remedied with external funding. On the other hand, the course of transitions can provide an opportunity for the creation of new businesses and new economic activities. Cape Verde's already long experience with regard to wind energy in particular, on the one hand, but also the existence of an important knowledge center (CERMI, ECREEE, higher education, etc.) can make this transition smoother, without major upheavals along the way.

Cape Verde is notable for the milestones it has already achieved in energy transitions, its strong political commitment, the involvement of international actors, and a focus on training and education. Understanding the reasons for this success is crucial in determining what steps are needed during this inevitably lengthy transition period for energy systems and economies. In addition to the significant investment required from developing countries, particularly SIDS, in their energy sectors—a challenge in itself given the scarcity of financial resources—it's also vital to ensure the education and training of a qualified workforce. This includes both higher and intermediate level professionals who will be equipped to handle the new processes inherent in modern energy systems. Cape Verde is taking this into consideration, both in offering higher education at university and also in the capacity it has shown to attract institutions such as ECREEE, which is the result of its integration into a regional organization of African countries (ECOWAS).

The evidence presented in this chapter suggests that the existence of expertise in developing countries can be a critical point for their sustainable development. If these countries, particularly the small island developing states that have more difficulties because they are small markets, do not fix this expertise, they end up creating another dependence on the outside, not on oil but on expertise in technology. Regional international cooperation can play a key role here in providing conditions for small countries to end their current dependence on oil without creating new dependencies. The diversification of international cooperation, not only in regional international cooperation, can be another relevant step, either by the greater possibility of opening the range of expertise to be developed in the small country, or by participating in larger markets and, therefore, having more opportunities and higher levels of efficiency. The case study of Cape Verde presents inspiring examples of cooperation initiatives that have produced tangible and meaningful results and built local capacities.

In the long run, the increased usage of renewable energy and the promotion of energy efficiency will substantially reduce oil and/or gas imports. This will free up financial resources that can be applied to other economic activities, including the development of the energy sector, or allocated towards other social sectors, such as health, education, or food provision. However, each country will face significant challenges in achieving these outcomes, and these challenges must be deemed acceptable in both economic and social terms. Addressing these challenges requires tailored approaches that consider each country's unique context while leveraging common opportunities to support sustainable energy transitions and socioeconomic development.

The UNOSSC (2021) concludes, "there is increasing recognition that South-South cooperation between Small Island Developing States (SIDS) and triangular cooperation (involving multilateral partners or the donor community) can accelerate the attainment of global international frameworks like the SAMOA Pathway and the SDGs." As a result, to advance in energy transitions, developing countries and their partners must adopt a comprehensive strategy that not only considers each country's unique circumstances and challenges, but also maximizes the potential of cooperative approaches. Doing so will not only accelerate socioeconomic development, but also foster a more resilient and sustainable future for all.

References

ALER. (2022). *Casos de Estudo: Energia renovável na Central de Dessalinização de Porto Novo, Cabo Verde*. https://www.lerenovaveis.org/pt/pesquisar/energia-renovavel-na-central-de-dessalinizacao-de-porto-novo-cabo-verde/

Alonso, J. A., Luiza Cortez, A., & Klasen, S. (2014). *LDC and other country groupings: How useful are current approaches to classify countries in a more heterogeneous developing world?* http://www.un.org/en/development/desa/papers/

Bass, S., & Dalal-Clayton, B. (1995). *Small island states and sustainable development: Strategic issues and experience SESA of REDD+ Strategy View project*. https://www.researchgate.net/publication/237111738

Câmara Comércio Sotavento. (n.d.). *Anglobal Incorporation*. Retrieved July 22, 2023, from https://www.ccs.org.cv/index.php/pt/empresas/anglobal

Cooperación Española. (2022). *El Pacto de los Alcaldes en África Subsahariana organiza el día CoM SSA enfocado en adaptación, previo a la #COP27*. https://www.cooperacionespanola.es/el-pacto-de-los-alcaldes-en-africa-subsahariana-organiza-el-dia-com-ssa-enfocado-en-adaptacion-previo-a-la-cop27/

ECREEE. (2015). *Centre for Renewable Energy and Industrial Maintenance of Cabo Verde Inaugurated*. http://www.ecreee.org/news/centre-renewable-energy-and-industrial-maintenance-cabo-verde-inaugurated

ECREEE. (2020). *Elaboração Do Programa Para O Corredor De Energia Limpa Da África Ocidental E Seu Quadro Institucional De Implementação*. http://www.ecreee.org/sites/default/files/procurment/attachments/2._tors_-_wacec_program_elaboration_pr.pdf

ECREEE. (n.d.-a). *Business plan 2011-2016*. http://www.ecreee.org/sites/default/files/documents/basic_page/ecreee_business_plan_final_design.pdf

ECREEE. (n.d.-b). *Overview of ECREEE*. Retrieved July 22, 2023, from http://www.ecreee.org/page/overview-ecreee

European Commission. (n.d.). *International Partnerships | Cabo Verde*. Retrieved July 22, 2023, from https://international-partnerships.ec.europa.eu/countries/cabo-verde_en

Expresso das Ilhas. (2019). *China está disposta a compartilhar as experiências bem-sucedidas com países como Cabo Verde*. https://expressodasilhas.cv/pais/2019/10/06/china-esta-disposta-a-compartilhar-as-experiencias-bem-sucedidas-com-paises-como-cabo-verde/65945

Governo da Guiné-Bissau. (n.d.). *No âmbito da implementação da Política para as Energias Renováveis da CEDEAO (EREP)*. https://www.aler-renovaveis.org/contents/files/080917_paner-gb_v71-final.pdf

Governo de Cabo Verde. (2018). *Plano Estratégico de Desenvolvimento Sustentável 2017/2021*. http://observatoriotabaco.cv/download/36/publicacoes/211/peds-2017-2021-versa%cc%83o-final.pdf

Governo de Cabo Verde. (2019). *Resolution No. 39/2019 approving the Master Plan of the Electric Sector 2018-2040*. https://www.fao.org/faolex/results/details/en/c/LEX-FAOC185115/

Governo de Cabo Verde. (2023). *Boa Vista International Conference of Partners 2023*. https://peds.gov.cv/caboverde4dev/wp-content/uploads/2023/04/ENERGY_2023v1.pdf

Governo de Cabo Verde. (n.d.-a). *Programa de Governo 2021-26*. https://www.governo.cv/documentos/programa-do-viii-governo-constitucional-da-ii-republica/

Governo de Cabo Verde. (n.d.-b). *The mid-Atlantic gateway to the world's economy | Energy sector*. Retrieved July 22, 2023, from http://www.ecowrex.org/sites/default/files/documents/projects/ennergy-sector-web_1.pdf

Herbert, S. (2019). *Development characteristics of Small Island Developing States*.

House, W. (2013). *Population and sustainable development of small island developing states: Challenges, progress made and outstanding issues*.

Instituto Camões. (n.d.). *Programas Estratégicos de Cooperação*. Retrieved July 22, 2023, from https://www.instituto-camoes.pt/activity/o-que-fazemos/cooperacao/atuacao/programamos/programa-estrategico-de-cooperacao

IRENA. (2019). *A new world: The geopolitics of the energy transformation*.

IRENA. (2022). *Sids Lighthouses Initiative Progress and way forward*. https://islands.irena.org/-/media/Files/IRENA/Agency/Publication/2022/Aug/IRENA_SIDS_LHI_Progress_Way_Forward_2022.ashx

Leandro, F. J., Martínez-Galán, E., & Gonçalves, P. (2023). *Portuguese-speaking Small Island developing states: The development journeys of Cabo Verde, São Tomé and Príncipe, and Timor-Leste*. Palgrave Macmillan.

Les Smart Grids. (2018, November 15). *Le Cap-Vert fait lui aussi le pari des énergies renouvelables*. https://les-smartgrids.fr/cap-vert-pari-energies-renouvelables/

OECD. (2018). *Making development co-operation work for small Island developing states*. OECD. https://doi.org/10.1787/9789264287648-en

RELOP. (n.d.-a). *Plano Estratégico 2022-2025*. https://relop.org/wp-content/uploads/2022/06/Plano-Estrategico-2022-2025.pdf

RELOP. (n.d.-b). *Sobre nós*. Retrieved July 22, 2023, from https://relop.org/a-relop/sobre-nos/

U.S. Department of Commerce. (2022). *Cabo Verde - Country Commercial Guide*.

UN News. (2020, September 29). *Cabo Verde quer chegar a 2030 com mais de 50% de energia renovável*. https://news.un.org/pt/story/2020/09/1727842

United Nations. (1993). *Report of the United Nations Conference on Environment and Development*. https://documents-dds-ny.un.org/doc/UNDOC/GEN/N92/836/55/PDF/N9283655.pdf?OpenElement

United Nations. (2018). *Accelerating SDG7 Achievement Policy Briefs in support of the first SDG7 review at the UN High-Level Political Forum 2018*. https://sustainabledevelopment.un.org/content/documents/18041SDG7_Policy_Brief.pdf

United Nations. (n.d.). *Decisions by topic: Small island developing states*. Retrieved July 22, 2023, from https://sustainabledevelopment.un.org/topics/sids/decisions

United Nations Office for South-South Cooperation. (2021). *Good Practices in South-South and Triangular Cooperation for Sustainable Development in SIDS | Advancing the SAMOA Pathway and Achieving Sustainable Recovery G*. https://www.un.org/ohrlls/sites/www.un.org.ohrlls/files/good-practices-in-sstc-for-sustainable-development-in-sids-web.pdf

UNOSSC. (2021). *Good practices in South-South and triangular cooperation for sustainable development in SIDS*. https://unctad.org/news/small-island-developing-states-face-uphill-battle-covid-19-recovery

World Bank. (2021). *World bank supports sustainable energy development in Cabo Verde*. https://www.worldbank.org/en/news/press-release/2021/12/08/world-bank-supports-sustainable-energy-development-in-cabo-verde

World Bank. (n.d.). *Cabo Verde: Development news, research, data*. Retrieved July 22, 2023, from https://www.worldbank.org/en/country/caboverde

João Simões is an Assistant Professor at City University of Macau, China. He earned his Ph.D. in Portuguese-speaking Countries Studies from the same university, a Master's degree in Chinese Studies from the University of Aveiro, Portugal, and a Bachelor's degree in Electrical and Computer Engineering from the University of Lisbon, Portugal. He co-edited the book "*Changing the Paradigm of Energy Geopolitics: Security, Resources, and Pathways in Light of Global Challenges*," published by Peter Lang Publishers in 2023.

SPRINGER NATURE

GPSR Compliance

The European Union's (EU) General Product Safety Regulation (GPSR) is a set of rules that requires consumer products to be safe and our obligations to ensure this.

If you have any concerns about our products, you can contact us on ProductSafety@springernature.com

In case Publisher is established outside the EU, the EU authorized representative is:

Springer Nature Customer Service Center GmbH
Europaplatz 3
69115 Heidelberg, Germany

The manufacturer's authorised representative in the EU is Springer
Nature Customer Service Centre GmbH, Europaplatz 3, 69115 Heidelberg,
Germany. If you have any concerns regarding our products, please
contact ProductSafety@springernature.com

Printed and bound by CPI Group (UK) Ltd, Croydon, CR0 4YY
30/09/2026
02225592-0005